...ndyford, Dublin 16
...1e 2078513 Fax 2959479
...1 library@imi.ie

Job Insecurity

Job Insecurity
Coping with Jobs at Risk

Jean Hartley, Dan Jacobson,
Bert Klandermans and
Tinka van Vuuren
with
Leonard Greenhalgh
and Robert Sutton

SAGE Publications
London ● Newbury Park ● New Delhi

First published 1991

SAGE Publications Ltd
6 Bonhill Street
London EC2A 4PU

SAGE Publications Inc
2455 Teller Road
Newbury Park, California 91320

SAGE Publications India Pvt Ltd
32, M-Block Market
Greater Kailash – I
New Delhi 110 048

British Library Cataloguing in Publication Data

Job insecurity: coping with jobs at risk.
  1. Job security
  I. Hartley, Jean *1953–*
  331.2596

  0-8039-8253-4

Library of Congress catalog card number 90-062744

Typeset by Mayhew Typesetting, Bristol, England
Printed in Great Britain by Billing and Sons Ltd, Worcester

Contents

Preface

The idea for engaging in collaborative research which ultimately led to the present book can be traced back to our participation in a symposium on job insecurity held at the West European Conference on the Psychology of Work and Organization in Aachen, West Germany, in the spring of 1985. During that symposium it became clear that we shared a common feeling that the area of job insecurity deserves considerably more conceptual and empirical work than has hitherto been conducted. We were convinced that this was an important research area and that we could explore it together. We felt that a concerted and systematic effort in teasing out the various components of job insecurity and how they affected three major levels of analysis – the individual level, the industrial-relations level and the organizational level – would be a timely contribution to the field.

A series of further meetings produced a research network which enabled us to co-ordinate our work while we carried out research simultaneously in our respective countries – Israel, the Netherlands and the UK. Our periodic meetings and the exchange of dozens of letters contributed to the cohesion of the programme for this book and provided important opportunities for cross-fertilization between individual perspectives. This procedure is time-consuming and expensive. It presupposes great frankness in discussion, as well as a desire on the part of each participant to understand the preoccupations of other participants. Of course, it has not been possible to meet all wishes, and compromises have been necessary regarding book content and details. On the whole, however, we feel that the separate but joint experience has been fruitful and enriching.

Although all of us have closely collaborated in the preparation of this book, neither in planning it nor in putting its several pieces together has there been an attempt to force individual perspectives into a common mould in the interest of some a-priori sense of theoretical or methodological integrity. Whatever may have been lost in the logic of organizational structure and in internal consistency has hopefully been fully compensated by the eclecticism that has resulted. The primary goal of this book is to open up the area from the theoretical point of view, to summarize our research and to offer implications for action. It is intended primarily for academic researchers, human-resource specialists, managers, trade union officials and consultants engaged in planning for the human and organizational consequences of uncertainty and job insecurity. The chapters in this volume do not purport to exploit all of the data that we have collected by the surveys. Rather, each chapter focuses on an aspect that has been of particular interest to us and that has a significant bearing on the job insecurity issue. Accordingly, also, some of the empirical chapters are based on quantitative data, while others rely on a qualitative approach. Because of the way we have organized the chapters, it is possible for researchers and practitioners alike to select and read chapters pertinent to their specific interests.

Chapter 1, the introductory chapter, begins with a brief analysis of recent objective developments in organizations' external and internal environments which may give rise to an increased level of experienced uncertainties among different categories of workers. We proceed by delineating the boundaries of our inquiry in terms of the population involved. We point out that our main interest in this book is in organizational employees for whom job insecurity involves a fundamental and involuntary *change* from a belief that one's position in the employing organization is safe, to a belief that it is not. We then demonstrate how most social psychological studies of job loss and unemployment have neglected job insecurity as a distinct phenomenon. This neglect stands out sharply given the growing concern about jobs in many organizations. Finally, in this chapter, we describe the specific job-insecurity context in the countries in which our surveys have been carried out, and provide details of the samples on which they are based.

The first part of Chapter 2 is devoted to a systematic analysis of the ways in which the job-insecurity experience differs from the job-loss experience. We argue that this distinction is important because the sources of stresses and tensions as job insecurity is encountered are not necessarily the same as when the job is lost. In the second part we discuss approaches to the conceptualization of job insecurity and review attempts to operationalize the job-insecurity construct. We then propose our own definition of job insecurity which served as the basis for the job-insecurity measure used in the studies described in this book.

Chapters 3, 4 and 5 address questions of what it means for individual employees to be working in an organization where jobs are at risk. In Chapter 3 a model is proposed for the analysis of individual responses to the job-insecurity experience. The model builds upon earlier theoretical work in the area, as well as on the literature on stress, causal attribution and participation in collective action. It hypothesizes that feelings of job insecurity are generated by factors that increase the perceived likelihood and the perceived severity of losing one's job. The model further attempts to map the response and coping patterns once job insecurity is experienced and suggests explanations for the different patterns that may be adopted by individuals. Chapters 4 and 5 review the empirical evidence in light of the model as provided by our studies in Israel, the Netherlands and the UK. Chapter 4 presents the findings on the determinants of the job-insecurity experience, that is, what makes employees feel insecure about the future of their jobs. Chapter 5 reports on the consequences of job insecurity in terms of psychological well-being and the different coping strategies that individual employees adopt. A key point is that workers in given organizations differ widely not only in the degree of job insecurity they experience, but also in their reactions to the experience.

The implications of job insecurity for the relations between management and workers both individually and collectively are treated in Chapters 6 and 7. Building on the insights of collective functioning immediately preceding, Chapter 6 sketches out why job insecurity is such an important, albeit neglected, area to consider in understanding changes in industrial relations. In particular, it points to the need to understand the subjective perceptions and interpretations of workers, since several outcomes from job insecurity are possible. It then goes on to examine power as an important aspect of union–management relations in this context. Chapter 7 explores the validity of these ideas in the case study of an organization in major change due to recession. The impact of job insecurity on industrial-relations attitudes and behaviour in this organization is delineated.

In Chapters 8 and 9 the focus shifts to the organizational level of analysis. Chapter 8 examines the influence of job insecurity on the propensity to leave the organization, on job involvement and job effort, and on resistance to change. The empirical evidence for each effect is briefly reviewed, then a model is presented that indicates their combined impact on organizational effectiveness. This system-level model posits a reciprocal relationship between job insecurity and organizational effectiveness with feedback loops that accelerate organizational decline. Chapter 9 builds on that approach and suggests strategies that organizations can adopt to cope with these dynamics and thus prevent or manage the destructive effects of job insecurity on organizational effectiveness. Both these chapters rely heavily on studies conducted in the US linking them to the perspective and evidence from Europe and Israel presented in earlier chapters of the book.

In Chapter 10, the concluding chapter, the major points made throughout this volume are integrated with emphasis on their practical implications. An attempt is made to speculate on the direction the problem is likely to take in a world of rapid and ever more difficult change. It is the authors' view that placing a high premium on employment security is an essential condition for a smooth adjustment to change in the future. However, employment security is not the same as job security. It does not imply a right to cling to a particular job forever. Such a right would indeed connote a kind of protectionism, a rigidity, on the part of both managers and workers, and an unwillingness to change. Employment security, as it is advocated here, results from the positive actions taken jointly by government, employers and unions to ensure that, in adjusting to major organizational change, the negative impact of job insecurity is avoided or minimized.

Chapter 8 of this book has been contributed by Leonard Greenhalgh and Robert Sutton, and Chapter 9 by Leonard Greenhalgh. Leonard Greenhalgh is Professor of Management in the Amos Tuck School of Business Administration at Dartmouth College, New Hampshire. His pioneering work in the areas of job insecurity and organizational retrenchment has provided important stimulation to our ideas and research. Robert Sutton is Associate Professor of Organizational Behavior in the Department of Industrial Engineering and Engineering Management at Stanford University, California. Professor Sutton is a widely recognized authority in the areas of organizational decline and death, and emotion and stress in organizations. Responsibility for the writing of all of the book's other chapters has been assigned to the four principal authors as indicated respectively in chapter titles. Each of the principal authors had full access to the data collected in the three participating countries, and chapter drafts have benefited from comments made by all of them.

Acknowledgements

The Israeli study described in this book was funded by the Golda Meir Institute for Social and Labour Research at Tel-Aviv University. The Dutch study was made possible by a grant from the Netherlands Organization for Scientific Research. The Economic and Social Research Council funded the British study while Jean Hartley was Senior Research Fellow at the Industrial Relations Research Unit, University of Warwick. In writing this volume, we have received assistance from several people. Gideon Gilotz, Jan van Gastel and Margaret Morgan assisted with the collection and analysis of data in Chapters 4, 5 and 7. Paul Marginson provided excellent comments on Chapters 6 and 7. Finally, we wish to thank our families and colleagues for the warmth and encouragement they provided throughout the project.

Dan Jacobson, Tel-Aviv   Bert Klandermans, Amsterdam

Jean Hartley, London   Tinka van Vuuren, Amsterdam

1

Mapping the Context

Dan Jacobson and Jean Hartley

Everybody is going around on pins and needles wondering if they'll be next. We are all reminded of what's happening every day. We have rows and rows of empty cubicles and desks. (An AT&T employee quoted in *Time*, 1987: 38)

Employees are running so scared that there is a whole culture that says don't make waves, don't take risks – just at a time when we need innovation. (A senior vice-president of a large corporation quoted in *Time*, 1989: 27)

Workers have a right to be upset and angry. They've been bought and sold and have seen their friends fired in large numbers. There is little bond in many companies between employers and workers anymore. (Chief economist for the AFL-CIO quoted in *Time*, 1989: 26)

These two *Time* magazine reports illustrate one of the most salient issues in many organizations. It involves the need to adapt to leaner times and to cut back on costs. This need reflects the increasingly frequent and often unpredictable swings in national and international economies, budget cuts in the public sector, and aggressive competition, deregulation and technological innovation. The 1989 *Time* report indicates a fairly long list of well-known corporations in Western countries considering a significant reduction of their workforce in the process of 'down-sizing' and restructuring. The same report presents results of an August 1989 survey based on a sample of American employed adults eighteen years of age or older in which 57 percent said that companies are less loyal to employees today than they were a decade ago. While 60 percent of the workers said they would prefer to stay in the job they have now, 50 percent said they expected to change jobs within the next five years. The same trend was noted in an earlier article which appeared in *Fortune*: 'Restructuring has put the final kibosh on traditional notions of corporate loyalty, whether of employee to employer ("As long as I do the work, my job will be secure, right?") or employer to employee ("As long as we take good, paternalistic care of you, you won't leave, right?")' (Kiechel, 1987: 81).

Fortune goes on to suggest the emergence of a 'New Employment Contract' contrasting sharply with the perception of the corporation in the 1950s and 1960s as a 'citadel of belongingness'. Thus, one of the chief features of the traditional 'psychological contract' between employee and organization (for example, Argyris, 1960; Levinson et al., 1962; Schein, 1980; Van Maanen, 1976) was its relative stability. It involved the employee's commitment to the organization in exchange for the latter providing a continuing source of current economic rewards and future security as well as (in many cases) psychological rewards. No longer, says *Fortune*:

> Hereinafter the company will be making no promises. The employee will assume full responsibility for his own career – for keeping his qualifications up to date, for getting himself moved to the next position at the right time, for salting away funds for retirement, and, most daunting of all, for achieving job satisfaction. (Kiechel, 1987: 81)

Indeed, the last fifteen years have witnessed high levels of organizational change brought about largely by recession and restructuring, though other influences have also been at work. The consequences of this trend for employees can be severe. Since changes in economic activity are reflected in the birth, alteration and death of private and public sector organizations, many people lose their jobs and unemployment increases in both numbers and duration.

But growing unemployment may simply be the tip of an iceberg; below lies uncertainty and anxiety for employees holding on (they hope) to their jobs in a sea of economic and organizational change. As restructuring occurs all round, and as redundancy and job loss abound, how long will their *own* job go on for? Also, as organizations try to reduce costs, there may be pressures on employees who remain at work to modify their jobs, accept different employment conditions, relocate. All this is likely to fuel job insecurity among those who remain in employment.

Economic and Organizational Restructuring as Antecedents of Job Insecurity

Since objective conditions feeding job insecurity have been so prevalent and seem likely to continue for some time, let us first examine some of them in greater detail. Awareness of the context of uncertainty is valuable in interpreting the subjective experience of job insecurity.

Recession

A major source of upheaval to jobs occurs during slumps in economic activity. Until recently many European countries have been in a recessionary epoch within a Kondratief 'long wave' which stems back from the period of the early 1970s and the oil crisis. Unemployment losses have been greater and more permanent than expected, and economic growth has been limited, following the long boom of the 1950s and 1960s. The period of 1979 to 1981 witnessed a particularly severe recession, resulting in widespread organizational closure and cutback. During the first part of the 1980s, world economic prospects were bleak. Many national economies appeared to be caught in a low-growth trap, their ageing and overmanned industries unable to compete successfully in world markets or even on their protected home turf. The period of growth which has followed during the last few years may represent a temporary boom in the shorter-run business cycle before many highly industrialized countries move back into predicted low growth and recession. Furthermore, in some large-scale and densely populated areas dominated by traditional industries the recessionary epoch continues to this day and is predicted to last. In these areas the threat to jobs is likely to continue towards the next century. The objective conditions for job insecurity, based in recession, may thus wax and wane, but will not disappear.

Restructuring, Mergers and Takeovers
While recession spells fewer jobs overall, restructuring involves fewer jobs in certain sectors as economic activity shifts from less to more profitable areas. Currently, many of the more developed countries are experiencing a shift from traditional manufacturing, as older, less profitable organizations decline or die, towards newer and service industries. While restructuring has always been part of a dynamic economy, it gained pace in the late 1970s. Restructuring has been exacerbated by the conditions of recession since there is a heightened concern by organizations with costs. Additionally, the greater internationalization of world markets since the 1970s has caused many organizations to consider a global rather than a national strategy. In Europe, the development of business in anticipation of the opening up of the European Community internal market in 1992 is already causing some restructuring and it is thought that much economic and organizational upheaval will happen as a result (Cecchini, 1988). More and more ambitious European companies, like US firms before them, are seeking greater global advantage by remodelling their operations. In the course of this process, some are streamlining by changing lines of activity which have shown low profitability and shedding marginal businesses acquired during earlier periods. Strong corporate earnings in North America and Europe over the past few years have allowed a considerable number of organizations to expand and reorganize through a pattern of mergers and acquisitions. The stock-market crash of 1987 helped stimulate the merger movement by bringing the price of many weaker companies within the reach of acquisitive activity. However, this amalgamation movement, while providing greater promise for jobs in certain areas, may create an increased amount of job insecurity in

other sectors of the economy and particularly among employees of takeover targets.

In addition to developments at the macro level, restructuring can also occur at the organizational level through the strategic choices of senior management about the deployment of resources. This happens for highly successful firms as well as for those in decline, as they shift capital, production and hence jobs not only between plants but between areas of the country and, increasingly, internationally (Blackaby, 1979; MacInnes, 1987; Massey and Meegan, 1982; Nickson and Gaffakin, 1984). The movement of organizations across regions, especially from union to non-union areas (Kochan et al., 1986; Massey and Meegan, 1982), and the transfer of jobs to the third world, increases the level of uncertainty. To the employees who may be at risk of job loss due to such redeployment of resources, job insecurity may be all too real, despite the healthy economic performance of their employing organization.

New Technologies

Newer technologies are another way for organizations to stay competitive and these too can spell employment change and uncertainty. In an international poll of employers (Leadbetter and Lloyd, 1987), 61 percent of the firms were reported as having introduced technologies that have affected their employment outlook. While early predictions of massive job loss through microchip technology were apparently exaggerated, there is reason to believe that dramatic changes in the employment scene will occur (Northcott et al., 1985). Whether introduced gradually or as part of wholesale change in the workplace, there is little doubt that microelectronic and other state of the art technologies do create concern over the continuance of employment. The introduction of such technologies may either eliminate jobs altogether, or raise the educational requirements to such a high level that workers with lesser schooling run the risk of losing their jobs.

Small Businesses

Another feature of restructuring in recent years has been the growth of small businesses. As pointed out in a recent report, there is reason to believe that 'jobs in small firms are not of the same quality as positions in larger firms' (OECD, 1985a: 79). The report indicates that the average job tenure in such firms is only 4.5 years compared to 9 years in a large firm. Moreover, the mortality rate of small businesses is relatively high. Consequently, quite apart from the generally lower levels of financial and fringe benefits, those employed in small organizations are at greater risk of job loss, and hence of some job insecurity.

Employment Flexibility

In their attempts to be competitive, many organizations have considered introducing employment flexibility, which according to recent reports is gathering momentum (ACAS, 1988; OECD, 1985b). This applies to both forms of flexibility: functional flexibility (where workers engage in a wider range of tasks) and numerical flexibility (where workers are hired on contracts which allow easier regulation of numbers by the employer). Numerical flexibility, in particular, has important implications for job insecurity. A systematic review of this form of flexibility has been provided recently by Pfeffer and Baron (1988). These writers describe the growing movement away from long-term employment relations towards increased 'externalization' of work – that is, diminished contractual attachments between employees and their work organizations. Major forms of this trend are shorter tenure, limited-duration contracts in a given organization, temporary and part-time work, sub-contracting, home-working and a general reduction in the organization's administrative responsibility for the employee. Good examples of this trend are networks of small entrepreneurs on contract for a large organization. Pfeffer and Baron (1988) cite American and British studies (for example, Sugarman, 1978; Atkinson and Meager, 1986) that illustrate the growing prevalence of two distinctive workforces within firms across various branches of the economy: the core (presumably permanent and relatively secure employees who are provided with opportunities for training and advancement), and the periphery (growing numbers of buffer employees who provide flexibility in the face of shifting workloads and who are given little training, offered few opportunities for advancement and given hardly any promise of job continuity). The emergence of these two workforces has been described as the 'Japanization' of employment (Brown, 1983b). A similar trend has been noted in multinational firms (Hodson and Kaufman, 1982).

Government Policies

Economic and organizational change has also been exacerbated in some countries by governmental policies, which eschew intervention through industrial policy (Leadbetter and Lloyd, 1987; MacInnes, 1987). Whereas in the 1970s governments showed a tendency to subsidize declining industries, in the 1980s they were much less inclined to do so. A number of governments have supported the drive for flexibility because of their belief in a market-driven economy. Support has, in some cases, consisted of the relaxation of employment legislation protecting workers from unfair dismissal in the belief that it makes the replacement of staff less costly (OECD, 1985b). The consequence for individual workers is to experience less protection from arbitrary management. Employers may also act more boldly in dismissing individuals or their union representatives in countries where the government espouses market forces (for example, Kochan et al., 1986; Lawler, 1986; Leighton, 1986). Such trends result in employment and trade union rights which give the worker less objective security.

A commitment to market forces has meant little economic and social cushioning of the massive organizational changes which have been taking place through recession and restructuring. And often the commitment to a non-interventionist stance is accompanied by a belief in curbing public spending. This has two effects on job insecurity. First, the cuts in public expenditure mean that jobs in the state sector are threatened as never before. Secondly, cuts in welfare and unemployment benefits make unemployment even less attractive and may increase the anxieties of those still in work, exacerbating feelings of insecurity.

Macro- and Micro-Level Implications
Overall, we have sketched a picture of economic and organizational change in industrialized countries in the late 1980s. The consequences for job insecurity are potentially serious – and likely to continue for some time into the future. In the face of the developments in the structuring of work arrangements there is reason to believe that employment may become considerably less stable for many segments of the workforce. Clearly, the broad public interest has a stake in these developments because it is affected – at a macro level – by what happens to the collective sum of employee–organization exchanges at a micro level. Although this is well beyond the scope of the present book, it would appear that society at large would have difficulties once decreased job security becomes pervasive across a large number of individuals and organizational situations. For example, this could have a negative impact on work motivation, willingness to invest and participate in occupational training programmes, welfare costs and so forth.

If, as has recently been suggested by Bluestone, Harrison and Clayton-Mathews (1986), deregulated sectors of the economy and even some still regulated sectors are undergoing a massive change in both the amount and the nature of uncertainties faced, then fixed capital of all forms, including personnel, may represent a potentially serious liability in the thinking of many employers. In that respect the current and anticipated trends may represent a renewed conflict between the interests of labour and capital. Pfeffer and Baron (1988) survey a long list of diverse advantages that diminished attachments between employees and employers are alleged to have for organizations. But they conspicuously neglect to point out the implications of the evolving trends for the individuals affected. While the existence of objective conditions for job insecurity does not guarantee its experience subjectively, the above consideration of major upheaval argues most cogently for a growing concern by employees regarding their employing organization's stability. However, contrary to the prevalent views among many employers, decreased job security among employees leads to a variety of adverse effects for the organization as well. This organizational perspective will be extensively discussed in Chapters 8 and 9. At this point suffice it to note that a policy study published recently by the Work in America Institute (Gutchess, 1985) forcefully argues that it is in the interests of corporate managers to regard their employees not only as a valuable resource precisely during times of change and restructuring, but also as a form of capital, human capital, to be protected and conserved if at all possible.

Whose Job Insecurity?

In the most general sense, job insecurity reflects a discrepancy between the level of security a person experiences and the level she or he might prefer. Seen in this broad perspective, it applies to several relatively large categories of workers who have little in common in other important respects. Perhaps the most vulnerable to job insecurity is the category of workers who belong to the secondary labour market (foreign workers; immigrants; ethnic minorities; older workers, including those who have retired from full-time jobs; and to a lesser extent women, particularly those with small children). Many in this category are employed as seasonal workers or as part-time or temporary help. Often they are leased from temporary-help service firms or labour contractors, and used as a protective organizational buffering mechanism (see Chapter 9). To a large extent, the total absence of security and stability of employment among members of this category reflects their peripheral position in the labour-market opportunity structure (Taubman and Wachter, 1986). It persists regardless of whether or not there are plans for cutbacks or retrenchment in the organization they happen to be working for at a given moment. Workers belonging to this segment of the workforce are also likely as a whole to be lower paid, to be non-union, with little access to pension plans and so forth (Taubman and Wachter, 1986). The grave social, economic and political implications of their marginal status in the labour market has long been a cornerstone of research on prejudice and discrimination in the industrialized world (for example, Castles, 1985; Doeringer and Piore, 1971; Quinn, 1979; Rogers, 1985; Semyonov and Levin-Epstein, 1987; Smith, 1974; Wallimann, 1984).

The second category of workers for whom job insecurity is by definition an integral part of the work experience consists of freelance workers; fixed-term appointments (for instance, in academic departments without the authorization to hire permanent faculty); or professional and technical staff working on a contract basis for higher pay (frequently, former permanent employees of the organization) who are hired with explicit understandings that employment will be of limited duration. For many of these workers job insecurity as a 'built-in' component of the work experience reflects the fact that no permanent jobs are available. But there is evidence that the idea of frequent changing of employers or of working on a freelance basis is gaining respectability among high-status occupations such as engineers, computer programmers and accounting professionals since it may offer convenience and flexibility in work schedules and location (for example, Appelbaum, 1985). Still, one should be cautious about attributing the acceptance of unstable workstyles to the genuine a-priori desires of all of those involved. As has been pointed out by Pfeffer and Baron (1988), workers who are only offered temporary employment may come to terms with this situation by 'choosing' not to desire a permanent attachment to a given employing organization.

The third category of workers who may be experiencing a considerable amount of job insecurity consists of newcomers to the organization who are still going through the induction stage. As pointed out by Van Maanen (1976), the encounter period for the newly recruited member is likely to be an extremely trying one. He or she enters what is often an alien territory, full of unforeseen surprises. Personal survival in the new organizational setting is clearly a major concern for the novice and he or she is undoubtedly more vulnerable to job insecurity than long-standing members of the organization (Jick and Greenhalgh, 1980). However, the risk of early withdrawal from the organization (both voluntary and involuntary) is but one aspect of the 'price of membership' (Schein, 1968) during the 'breaking-in' period (Van Maanen, 1976).

The feature that is common to the first two of the foregoing categories of workers is that the discrepancy between the level of security experienced and the level that is desired is an ongoing, inseparable part of their worklife. Members of these groups usually have a relatively stable set of beliefs about the labour market and their prospects in it. By virtue of either their marginal status in that market, their occupational speciality or their own preference, they are placed in a continually insecure position. For the third category, that of the newcomers to the organization, job insecurity is often only a stage in the employment process willingly entered into. Here the discrepancy between the level of security that is experienced and the level that is desired is endured temporarily with the objective of eventual permanent employment clearly in mind.

The focus of the present book, however, is on change rather than on discrepancy alone. Hence our interest in a fourth category of workers who are experiencing job insecurity. The major element that distinguishes the job-insecurity experience in this category from that of the previous three is that it involves a fundamental and involuntary *change* in the worker's set of beliefs about the environment or about his or her employing organization and his or her place in it. Specifically, the change is from a belief that one's position in the organization is safe to a belief that it is not. In other words, the change in beliefs sets in *within* the organization as it (rather than the person) moves from what has been characterized by Pettigrew (1983) as a 'rich' to a 'poor' environment. The 'rich' environment is one that 'is featured by a strong emphasis on expansion and growth' (Pettigrew, 1983: 105). In the 'rich' environment, as Pettigrew (1983: 105) puts it, career concerns are centred on 'still larger empires and faster promotion'. Conversely, the 'poor' environment is one in which 'the rhetoric of finance, economics, and accounting are linked to strongly articulated values about efficiency, and all these are harnessed to the new pre-occupation of resource management while career concerns move to personal survival' (Pettigrew, 1983: 105–6).

The central interest of the present book, then, is in the growing number of workers who are experiencing a transformation in their beliefs regarding the durability of their present employment relationship, rather than in those equally vulnerable (but hitherto more thoroughly researched) members of the labour force for whom job insecurity is an ever-present threat due to their demographic background, the nature of their work or their specific employment status. To be sure, the exclusion from the frame of reference of this book of the three categories of workers described in the first part of this section does not by any means imply that their experience of job insecurity is less severe. They are put on one side because, by definition, job continuity in the same given organization is not an integral part of their set of expectations in the first place, or cannot be taken for granted at the present stage of the employment relationship.

It follows that the term 'job insecurity' as used here is limited to permanent employees, who are past the organizational induction stage. Threats to job security in the case of these employees are of two types: a threat to the loss of the job regardless of the job-holder – through, for example, retrenchment, mergers, restructuring or the introduction of advanced technologies. Alternatively, job insecurity may arise through the loss or erosion of employment rights, for example, through changed contracts of employment. In the latter case, the job continues but the job-holder is vulnerable. As indicated previously, both types of insecurity can be a result of economic, legislative and organizational change.

For members of this category in the labour force, the work environment consists of a set of structured employment relationships, within the organizational internal control system, embodying an integrated set of rules. These rules may be formal (as in unionized and bureaucratic organizations) and informal in the form of a 'psychological contract' (for example, Argyris, 1960; Levinson et al., 1962; Schein, 1980; Van Maanen, 1976). They explicitly or implicitly cover the content and rewards attached to each job, the organizational structure that ties the jobs together, entry requirements for new hires, exit procedures, career and promotion opportunities, and so on. For workers thus employed, the expectation of job continuity in a given organization generally is a major component of the psychological contract that governs the on-going transaction between them and their employer. Hence, the threatened disruption of that continuity constitutes a major breach of the psychological contract and consequently, in some cases, a serious personal crisis.

Past Research in the Area

While job insecurity is presumably an important issue whenever an organization is faced with hard times or major change, and is likely to become an increasingly important issue in the next few years, it has captured a fairly limited interest from scholars. Attempts at both developing a theoretical framework and conducting research to explain and predict variance in employee reactions to uncertainties in a given organizational context are still rudimentary (Greenhalgh and Rosenblatt, 1984). Probably a major reason for this neglect is that, although job insecurity is a phenomenon critical to those who work in organizations and manage them, it is less amenable to empirical research. Job insecurity is, of course, a highly individualized and sensitive topic. Because of its sensitivity and highly emotional overtones, many people are reluctant to become involved in its study. Indeed, the same problem was repeatedly encountered in the studies described in this book. Typically, organizations undergoing difficult times are extremely hesitant when asked to permit the gathering of data on such an emotionally 'hot' topic (see, for example, Sutton and Schurman, 1985). This is quite understandable since research efforts focused on job insecurity may in themselves generate anxiety. However, the result is that the pervasive organizational phenomenon of low job security has remained relatively under-researched.

This relative omission of job-insecurity research in the literature stands out sharply if one considers the importance attributed to job *security* in traditional industrial/organizational psychology. Not surprisingly, the emphasis on job security goes back to the Great Depression years of the 1930s. For example, Chant (1932) reports that managers indicated 'steady work' as one of the twelve most important factors in the evaluation of any job. Similarly, in a study of union and non-union employees interviewed between 1928 and 1934, 65 percent of union workers and 93 percent of non-union workers ranked 'a steady job' as the second most important management policy (Hersey, 1936). More recently, in line with the growing social psychological concern with job satisfaction during the 1940s, 1950s and 1960s, job security was intuitively considered vital to worker motivation. Maslow's (1943; 1954) popular motivational theory, which postulates that employees' needs are hierarchically arranged and become activated in a prepotent manner, is a good case in point. In his classic need hierarchy, when the basic physiological needs are satisfied, the safety needs become the most important. McGregor (1960: 37) has aptly summarized the potency of the safety needs in the organizational context: 'Arbitrary management actions, behavior which arouses uncertainty with respect to continued employment or which reflects favoritism or discrimination, unpredictable administration of policy – these can be powerful motivators of the safety needs in the employment relationship at every level, from worker to vice president.' Job security has been almost universally recognized as a major extrinsic component of job satisfaction. The dual-factor theory of motivation (Herzberg, 1966; 1968) was a theoretical cornerstone, albeit controversial, in that regard. In Herzberg's (1966) formulation, job security is not a motivator providing satisfaction with one's job. Rather, it is classified as a 'hygiene' factor which, when absent, could lead to job dissatisfaction. Other researchers, however, concluded that job security may act as a source of both satisfaction and dissatisfaction (Burke, 1966; Graen, 1966; Kornhauser, 1965). They also found that security could serve as both a motivation to stay with an organization and as a reason for leaving (Friedlander and Walton, 1964).

The importance attributed to job security over the years in relation to other job characteristics was persuasively demonstrated in a series of studies between 1946 and 1975 (Jurgensen, 1978) where job applicants were asked to rank the relative importance of ten job characteristics. Data were broken down into five-year intervals. In four of these intervals (1946–65), job security ranked first in importance. From 1966 until 1975 job security dropped to second place behind 'type of work'. Males tended to place greater emphasis on security than females. Job security was shown to increase in rank as a function of age over the study period, and decreases in security rankings were negatively related to educational level. The importance attached to security varied with previous occupations held by the applicants. Jurgensen (1978) admits that the implication of his data might be limited by the focus on job applicants (as opposed to job incumbents), by the type of corporation from which applicants were drawn (public utility), and by the specific geographic area (Minneapolis, Minnesota). However, given the large sample size (57,000 applicants over a period of thirty years), and the time-series nature of the studies, his data should not be ignored.

Moreover, other studies regarding the importance of job characteristics tended to corroborate the finding that job security is of major importance to workers. Differences between the sexes, at least in the US, are inconsistent and seem to have decreased markedly (or vanished completely) in recent years (Brief and Oliver, 1976; Brief et al., 1977, Schuler, 1975). Lahey (1984), in her extensive review of the job-security literature, suggests that the reason for this recent change may be a function of controlling for job level, since security is more important at lower job levels than at higher levels. However, this job-level difference, according to Lahey (1984), does not imply that job security is not important at higher occupational levels, only that it may be relatively less important than other needs.

Lahey (1984) points out that in spite of the documented importance of job security in people's work-related responses, previous literature has been slow to investigate the meaning of the construct. The lack of a sound theoretical base has led to confusion and inadequacies in both conceptualization and measurement. She cites a long list of studies in which job security has been inconsistently conceptualized as an attitude, a motivator, a reward or incentive, a 'need', and as a part of the organizational context.

Given the conflicting interpretations of job security in the literature, Lahey (1984: 164) defined job security generally as 'an expectation that employees would be allowed to continue in their present jobs, provided that job performance meets (or exceeds) company standards'. Her own study was not intended to measure the importance of security, nor was it designed to tap workers' affective responses with respect to their level of job security. Using factor-analytic procedures on the basis of survey data derived from 487 employees of three service, manufacturing and civil-service organizations, she developed a measure of job security which incorporated, somewhat questionably, antecedents of security into a multi-dimensional construct consisting of five different dimensions. These dimensions, she claims, cover aspects of the social and financial climate of an organization, as well as specific characteristics of individual employees and their jobs. They include Company Concern for the Individual, Job Performance, Company Growth and Stability, Job Permanence and Individual Commitment. Lahey (1984) reports that the overall job-security scale based on individual dimension scores accurately reflected differences in working conditions within the organizations studied, thus supporting the validity of her scale. From a more practical point of view, however, it would appear that one of the major problems with Lahey's (1984) measure lies in its lack of parsimoniousness.

Given the importance of job security whatever the exact meaning of the construct, it is not surprising that most existing research on job *insecurity* has focused on its negative consequences. This research deals primarily with the effects of the job insecurity experience on psychological well-being (for example, Davy et al., 1988; Depolo and Sarchielli, 1985), on group processes (for example, Krantz, 1985) and on various indicators of organizational effectiveness (Cameron et al., 1987). The latter category includes studies on the effects of job insecurity on productivity and turnover (Greenhalgh, 1982; Greenhalgh and Jick, 1979; Sutton, 1983), commitment and loyalty (Staw et al., 1981), union–management relations (Hartley, 1985), and the 'survivors'' esprit de corps (Brockner et al., 1985; Brockner et al., 1986; Davy et al., 1988; Greenhalgh, 1985). In general, these studies, which are reviewed extensively in Chapter 8, suggest that lower job security enhances psychological withdrawal from the job (that is, decreased organizational commitment and job satisfaction) and behavioural withdrawal (that is, absenteeism and quitting). Such reactions have been found, in particular, among 'survivors' who witnessed job loss among peers and who felt that their own jobs were threatened.

The intrapsychic level of the job-insecurity experience is even less well researched than the organizationally related effects. Briefly, the literature suggests that anticipation of job loss generates personal distress and sometimes even a grief reaction akin to that of anticipated bereavement (Greenhalgh, 1985; Strange, 1977). On the other hand, at least two studies (Kinicki, 1985; Swinburne, 1981) report – albeit on the basis of retrospective interviews with people who actually lost their jobs – that forewarning and expectation of plant closing in fact reduced negative affective responses because it encouraged the individual to better prepare and cope with the anticipated stressful life event. A somewhat more complex picture of both positive and negative aspects of the job insecurity experience was provided by Jacobson's (1987) content analysis of statements made by Israeli public sector employees confronted with a government decision on workforce cutbacks. This study revealed the coexistence of three seemingly contradictory job-insecurity-induced attitude clusters: (1) an attributive attitude comprising suspicion, external anger and self-blame; (2) a surrender attitude composed of feelings of unchangeability and loss of control resulting in demoralization, helplessness and stress reactions; and (3) a coping attitude consisting of a generally optimistic belief that the problems posed by job insecurity can be resolved.

The systematic measurement of job insecurity is only just beginning to emerge. The first theory-based perspective has been provided by Greenhalgh and Rosenblatt (1984), with some refinements by Jacobson (1985) and Brown-Johnson (1987). An empirical examination of the validity of the multiple measure derived from Greenhalgh and Rosenblatt's (1984) perspective has recently been attempted by Ashford et al. (1987). This and the theoretical rationale behind it will be discussed more extensively in the next chapter. Most other studies have measured job insecurity using single indicators rather than multiple indicators. In some cases such indicators were included in a battery of other potential job-related stressors (for example, Caplan et al., 1980; Johnson et al., 1984).

By and large, however, job insecurity has rarely been given a prominent focus in the literature. Usually, it has won only scant attention as a subdomain of the vast research literature dealing with the social psychological dynamics of involuntary job loss and unemployment (for recent critical reviews of this literature see DeFrank and Ivancevich, 1986; Fryer and Payne, 1986; Hartley and Fryer, 1984; Kaufman, 1982; Kelvin and Jarrett, 1985; Leana and Ivancevich, 1987; O'Brien, 1986; Warr, 1987). This literature provides us with considerable knowledge of the social factors and the cognitive and perceptual devices that function to shape the different meanings attributed to job *loss* at or following its occurrence. It has also greatly enriched our understanding of the coping repertoires and coping efficacy of individuals struggling with its stressful effects. The effects of future-oriented appraisals of the job-loss threat have, however, been largely ignored. A good case in point in this respect is Warr's (1984) analysis of the transition from the 'employed' role into the 'unemployed' role. In describing the removal of features of the earlier role as a result of job loss and their replacement with features associated with the second role, Warr (1984) appears to imply a clearcut dichotomy with no intermediate state between the two. As indicated previously, this reflects a characteristic neglect in the literature of job insecurity as a distinct phenomenon.

Job Insecurity in the Countries Participating in This Research

Having set the scene for the need to study job insecurity, we now turn to a brief description of the context of job insecurity in each of the three participating countries. In addition, we present some comments relating to the US job-insecurity scene which provide the background for Chapters 8 and 9. We follow that with some details of the research samples.

Israel

Since the establishment of the State of Israel in 1948, all governments have conducted an explicit policy of minimizing as far as possible the level of unemployment. This policy was deeply rooted in the predominant Zionist–socialist ideology which emphasized the return to productive labour of Jews immigrating to Israel and which stressed the need to provide employment as a necessary means for encouraging them to make Israel their home. Given the continuous threat to its security and the numerous potential conflict areas in Israeli society (on the national level: Jews versus Arabs; on the ethnic level: Oriental versus Western Jews; on the level of value systems: the politically strong religious minority versus the secular majority), economic and class tensions due to high unemployment have always been seen as dangerous additional burdens that Israeli society just cannot bear. Due to this policy, and aided by highly centralized governmental control and the powerful General Federation of Labour (Histadrut), unemployment until the mid-1980s rarely constituted a social stressor in Israeli society. Moreover, a myth of 'total job security' has become quite widespread and Israelis, by and large, have grown to take job continuity for granted even during periods of economic or political upheaval and a three-digit inflation.

To a considerable extent this state of affairs has changed dramatically since the summer of 1985 in direct consequence of the government's drastic anti-inflationary moves. These moves resulted in substantial cutbacks in the public sector. In addition, there emerged a widespread trend towards retrenchment and a growing number of closures both in the large trade-union-owned sector and in the private sector. In many cases, Histadrut, which encompasses most of Israel's trade unions, was compelled to agree to the loss of some jobs to save others. It has also supported painful restructuring efforts, hoping thereby to prevent closures. Reluctantly, it even had to give in to massive dismissals in some of its own large-scale industrial concerns. Prominent among the government's moves was the decision (never fully carried out) to seek the dismissal of 3 percent of the entire public sector workforce. Over 40 percent of all employed people in Israel are employed in this sector, from which the Israeli sample was drawn.

While the overall unemployment rate in Israel has increased in the late 1980s (hovering between 6.5 and 8 percent of the civilian work-force in 1988–9), it is still lower than in the two European countries represented in this study. Nevertheless, the myth of virtually guaranteed job security has been shattered. Many Israelis in recent years were confronted for the first time in their lives, and quite unexpectedly, with a potential threat to job continuity. Intensive and on-going coverage by the media, plus the small size of the country, contributed to the salience of job insecurity as an issue affecting the real-life experiences of many Israelis since the mid-1980s.

The Netherlands
During the first decade following the Second World War, the Nether-lands adopted a policy of general employment security. Unemployment rates were very low, and the level of job security was very high. Employers' associations, government and unions collaborated in the reconstruction of the Dutch economy in which a policy of general

employment was coupled with union support for a relatively moderate wage level.

In the long run, and in view of developments in other European countries, this policy of low wages to guarantee employment became unsustainable. In the 1960s, wages in the Netherlands increased towards European levels. The rapidly expanding economy of that decade was able to absorb these wage increases, but when during the 1970s a worldwide recession set in, both wages and employment security came under growing pressure. Unemployment rates shot up to a peak of 17 percent in the beginning of the 1980s, dropping back to approximately 12 percent in 1988. This was accompanied by a growing trend towards organizational restructuring, elimination of jobs in many companies and company closures. The job situation in the Netherlands was exacerbated even further by the Dutch government's attempts to reduce budget deficits. Consequently, a large number of jobs in the public sector were also lost.

As a direct result of these processes the position of the trade unions weakened considerably. Membership fell from almost 40 percent of the total workforce in the 1960s to less than 30 percent in 1988, and because of the high unemployment rate the bargaining position of the unions deteriorated. Although Dutch employers are legally required to consult workers' councils prior to restructuring and cutbacks, this was insufficient to dissuade employers from pushing through major restructuring plans resulting in massive dismissals. Still, in comparison with some other European countries and the USA, unions in the Netherlands remained more influential both on the national level – through regular consultations with employers' associations and government representatives – and on the local level – through the workers' councils. This has been demonstrated, for example, by their success in negotiating the shortening of the workweek from forty to thirty-eight hours.

United Kingdom

In the UK, after the Second World War, there had been a gradual strengthening of job security through collective bargaining in the period of relatively full employment of the late 1950s onwards. The UK's voluntarist tradition of employment conditions, however, meant that some sectors enjoyed reasonable levels of security, while others were characterized by high levels of uncertainty and change with low levels of protection for workers. In this period, job security, in a buoyant labour market, was not a priority for trade union demands.

However, despite the availability of jobs in the 1960s and early 1970s, the UK's industrial performance was declining (Blackaby, 1979; MacInnes, 1987). An attempt was made to increase the speed of industrial change with legislation to provide financial cushioning for those who became redundant (the Redundancy Payments Act 1967, amended in 1969). This caused some uncertainties for trade unionists and workers, who felt the contradictory pressures of resisting job losses, or negotiating for compensation above the statutory minimum.

Unemployment began to rise steadily from 1976 and then steeply from 1979, with job insecurity a corollary of the growing problems of recession and restructuring. The official male unemployment rate in 1979 was 4.2 percent but rose to 8.1 percent in 1981. It then continued to rise even with the onset of recovery in the economy in late 1982 although beginning to fall slightly by late 1987, with 7.1 percent being recorded in early 1988. However, precise comparison of figures is difficult because of complex changes in official statistics in the decade of Conservative government.

The collapse in employment in the 1980s was the worst ever experienced in the UK – worse than the famed years of economic difficulty in the 1920s and 1930s. The grim employment figures in the UK can be explained by the cyclical decline affecting all industrialized countries after 1979, but the severity in the UK was affected even further by the government's monetary and fiscal policies, which severely exacerbated industrial and employment problems. The government believed strongly in a non-interventionist stance on industrial affairs. It also enacted several pieces of legislation which weakened trade unions, made dismissal and job loss easier, and savagely curtailed the social security and unemployment insurance for those who lost their jobs. For those people anxious about the continuance of their jobs, there was plenty to be worried about as the living standards of those who became unemployed became widely divergent from those still in jobs.

The negative effects of restructuring on unemployment were especially marked for the West Midlands, an area of traditional manufacturing and the location of the UK study described in this book. For a manual worker living and working in that region, job security had been seriously affected by the deep cutbacks in output and employment in the trough of 1981. Although recovery began to take effect in 1982, redundancies were still at a higher level than in 1979, at the onset of recession, and the number of manufacturing jobs has failed to increase. Overall, in recent years, the rate of decline in the West Midlands has slowed but not ceased.

United States

Job security has always been a more tenuous affair in the US compared to Europe. Employment legislation gives the worker less protection against involuntary dismissal, and the financial compensation for many who lose their jobs is considerably less than in Europe.

The US led the recession that affected the world economy from 1979 – but was also the first major country to experience the recovery in demand. Unemployment rates increased from 5.8 percent in 1979 to 7.4 percent in 1981, and peaked at 9.5 percent in 1982 and 1983. After that, the unemployment rates have gradually declined. Overall, unemployment rates (standardized), while reaching a higher peak in the recession than some of the OECD countries, have also dropped more. Currently, joblessness in the US is at its lowest level since 1975, and the unemployment rate dropped to 5.3 percent in mid-1988. In contrast to the European experience, the American economy's expansion since 1983 has created more than fifteen million new jobs that numerous companies find difficult to fill. One of the main causes for this trend is the low birth rate in the late 1960s. This means that there are fewer teenagers and college-age students available today to take on the increasing number of jobs available in the service industry. To be sure, in many cases there are enough applicants, but far too few of them have the basic skills – reading, writing and arithmetic – to handle jobs in an economy that increasingly runs on technology and information. For the moment, then, the shortage of suitable candidates for available jobs has put no brake on the overall US economy. But if the supply grows tighter and forces employers to pay ever steeper wages, the situation could help spark a resurgence of inflation followed by yet another recessionary cycle. This development could be exacerbated even further by the huge US budget deficit.

The concern with job security in the US manifested itself in some of the demands in collective bargaining. 'Give-back' bargaining started to be a phenomenon of the recessionary years of the early 1980s, with unions agreeing to wage cuts and job flexibility in exchange for some limited guarantees about job tenure. In addition to the recessionary influences and the subsequent widespread trend towards restructuring in the US, job insecurity was affected by the move to union-free environments, mainly in the sun-belt, right-to-work states of the south-west (Kochan et al., 1986).

The Data Base for This Book

The findings reported in this book are based on structured or semi-structured interviews and questionnaire completion with members of organizations threatened with cutback crises in Israel, the Netherlands and the UK. The observations relating to the US presented in Chapters 8 and 9 are based on previous research conducted by those chapters' authors. Since the workforces in the Dutch and the UK organizations were composed primarily of men, there were insufficient data to base any analyses on gender.

Certain parts of the questionnaires used in the studies were developed jointly by the investigators from the three countries and were comparable in content and often in wording. They explored the organizational changes occurring, the perceptions of and feelings about job insecurity and its causes, and the industrial relations within the workplace. No attempt was made to force exact similarities in wording, however, or to translate questions literally. Differences in national style, in ways of talking and approach, were reflected in each questionnaire. In each questionnaire some questions explored individual investigators' interests and preoccupations. In order to ensure some comparability between countries, all concepts and variables were defined broadly in the same way, both in the question-naires and in the tabulations. The investigators meeting together discussed coding of all questions and adopted standardized coding methods. The same procedure was followed in planning the presenta-tion of specific data.

The Israeli Study
As indicated before, for employees in the Israeli public sector the potential threat to job continuity had first become a salient issue follow-ing the government's decision, in July 1985, to seek the dismissal of 3 percent of the entire workforce in that sector. The timing of the Israeli study was therefore quite crucial. Respondents were interviewed less than a fortnight after the government's decision had been announced, receiving widespread coverage in the media. It was expected that most, if not all, respondents were acutely aware of the decision. None, however, on a personal basis had at the time of the study received any formal or informal indication of eventual dismissal.

The sampling method used in the Israeli study was that of 'organization-probability sampling', a method which gives each employee in the organization an equal chance of being located and thus participating in the research. However, the three highest levels in the Ministry's hierarchy were not included in the sample. The only other constraint on selection was the respondents' relative fluency in Hebrew (that is, excluding employees who had only recently emigrated to Israel). The original sample consisted of 258 white-collar employees in a government ministry who agreed to participate in the study after being personally contacted by the Israeli researchers. The Ministry's top officials granted permission to conduct the interviews during office hours with the anonymity of the participants fully guaranteed. The agreement and co-operation of the local workers' council was also gained and made known to all Ministry employees.

The interview stage produced 233 usable questionnaires (a 90.3 percent response rate). The mean age of the respondents was 38.3. Almost two-thirds (63.5 percent) were Israeli born, 45.5 percent had between nine and twelve years of formal education and 54.5 percent thirteen years and more. The vast majority (85.8 percent) had a tenured position in the civil service. The mean length of service was 10.8

years. All respondents belonged to the civil-service union which is part of Histadrut.

The Dutch Study

The study in the Netherlands was based on structured interviews held in three organizations. All three organizations have gone through significant changes recently.

The first organization, an engineering company, was restructured in 1986 for the third time since the 1970s, due to lack of demand for the company's products. In this most recent change, 25 percent of the workforce were dismissed. In June 1987 interviews were conducted among the survivors. At the time of the interviews the company's future appeared to be stronger than before. Productivity increased, morale improved, the company changed ownership and its managerial philosophy transformed from product-centred to market-centred.

The future of the second organization, a shipyard, depended on its ability to obtain new orders for its two main divisions – shipbuilding and ship-repair. In 1984 the shipyard had already sustained a cutback of 200 employees from its original workforce of 1900. Two years later, in 1986, orders were obtained for the repair division, but the shipbuilding division continued to fail. The situation worsened in 1987 with no new orders for shipbuilding and growing underemployment in the repair division as well. When interviews were held, in September 1987, it was clear to employees that the shipyard desperately needed new orders before 1988 to avoid further decline. However, the ability to obtain new orders partly depended on the government, which was reluctant to increase the shipyard subsidies to assist them in their intense competition with other shipyards.

The third organization produced electrotechnical equipment. It went through two successive reorganizations in the early 1980s, continuing into 1983 with the firm constantly shifting to new products, changing customers and moving towards further automation. The interviews that were held in 1987 took place in one of the organization's new sections, where many employees had participated in retraining courses for using new computerized machinery. However, one month prior to the interviews management unexpectedly announced its intention to dismiss less qualified workers, while there were still vacancies for highly qualified workers. The reason given for the planned cutback in personnel was automation. The employees marked for dismissal were not included in the sample.

The sampling procedure in the Dutch study differed according to the extent to which each of the three participating organizations was affected by restructuring and cutbacks. If only one major sector of a company was affected, a probability sample from the workforce of that sector alone was drawn. If more than one sector was affected, a stratified sample was drawn so that each sector would be represented accordingly. All in all, 311 employees in the three organizations (72 percent of the original sample) were interviewed in their homes. Of these, 75 were employees of the engineering company, 106 were working for the shipyard and the remaining 130 were employed by the electrotechnical equipment company. Respondents were approached following the granting of permission by the management of each participating organization and after the local workers' council as well as the local trade unions had agreed to co-operate.

The mean age for all respondents in the three Dutch organizations was 40.1 years, and their mean length of service was 16.7 years. A small minority of the respondents (8 percent) were nationals of other countries. Over half (51.3 percent) had an above-basic level of vocational education. There was a difference between the three participating organizations in the level of unionization: the shipyard had the highest level of unionization (72.6 percent), followed by the engineering company (45.3 percent) and the electrotechnical equipment company (29.2 percent). The unionization level for the entire sample was 47.9 percent.

The UK Study
The UK case material was drawn from a manufacturing company with three plants in the same West Midlands town. The company manufactured vehicles in small-batch or one-off production. It employed a skilled workforce on a range of engineering activities associated with production.

The company had experienced considerable decline in both employment and output. Having enjoyed a reputation for being a leader in its specialized product, it had suffered badly from the onset of recession in 1979 and from increased competition from European and Japanese firms. By 1984 output was down to less than half the level five years previously, and the workforce had been reduced by nearly half, from around 2000 to 1250. There had been three main waves of redundancy between 1979 and 1983, and at the time of the study (1985–6) the opportunity to leave the company on redundancy terms was still open to employees, although the drive to reduce workforce numbers had eased. But the company remained in a precarious position due to its reduced output, market position and financial state. In fact, eighteen months after the study, the company went into receivership through financial insolvency. The company also suffered organizational problems through its having been sold from a large multi-plant enterprise to exist as a single company. As a result it lost much of the financial cushioning, especially cash flow, which it had previously enjoyed.

The questionnaire asking about job insecurity and industrial relations was distributed to members of the workforce. There were 137 replies, which represented a response rate of 42 percent. The questionnaire sample had a mean age of 44.5 years, and an average seniority in the company of fifteen years. Most of the sample were male (92 percent). They were all members of one of two unions (an informal closed shop existed). Their jobs were skilled or semi-skilled engineering jobs, predominantly in fitting and machining. As well as the questionnaire data, information was collected by just over 100 interviews with managers, trade union representatives, union full-time officials and additional members of the manual workforce. In addition, the researcher spent a six-week period of observation in the plant and in this way attended a variety of management, union and union–management meetings. Documentary material was also used.

2

The Conceptual Approach to Job Insecurity

Dan Jacobson

The mere statement that widespread apprehension about jobs exists and may be growing is of little value unless we find ways to present it as a clear and distinct experience. So a preliminary step towards better understanding of the job-insecurity experience is the recognition that there is no one-to-one correspondence between it and other employment-related crises and in particular with the job-loss experience. It has its own particular antecedents and is manifested in actions and attitudes. To be sure, in some respects job insecurity is a harbinger of the process of job loss. Earlier writers (for example, Hartley and Cooper, 1976; Joelson and Wahlquist, 1987) have begun to describe pre- and post-job-loss stages. We have built on, but modified, their analyses of stages and our representation of a three-stage sequence is given in Figure 2.1. Job insecurity is the first of these stages. It is associated with planned but unannounced redundancies. During this stage employees are unsure who (if anyone) will be forced to leave the organization. Joelson and Wahlquist (1987: 179) characterize the job-insecurity stage (the 'anticipatory phase') in the following terms:

> It is a period of agony of varying strength. Rumours about possible decisions and actions are circulating. Reliable information is not available. You have to decide whether you should try to look for another job or not. Sometimes you have too little to do. You hover between hope and despair. The trade union tries to get as many as possible active in the struggle to save jobs and to make the unemployment problem a concern for the entire working group and not the individual person. For the individual this phase is characterized by the threat of impending unemployment.

However, viewing job insecurity as a stage in a sequence does not mean that it is necessarily followed by job loss and unemployment, because the populations affected by job insecurity and job loss are only partly overlapping. In fact, the population experiencing job insecurity is in most cases considerably larger than the number of workers who lose their jobs. Moreover, an attempt to equate the impact of job loss on job losers with uncertainty about future job retention is not supported in the literature that relates to other catastrophic events that challenge adaptive abilities such as natural disasters (Baker and Chapman, 1962; Erikson, 1976; Janney et al., 1977; Sims and Baumann,

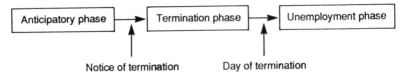

Figure 2.1   *Phases of the unemployment process (adapted from Joelson and Wahlquist, 1987: 180)*

1974), war-related events (Klein, 1974; Milgram, 1986), relocation (Liberman and Tobin, 1983), illness (Mechanic, 1978), divorce (Kressel, 1986), or significant loss (Parkes, 1972; Paulay, 1986). This literature as well as available typologies of life crises (for example, Berren et al., 1986) suggests that the significance of the encounter is appraised differently at various stages and calls for different modes of coping. Thus, the period of anticipation, the period of impact and the post-impact period (in our case the job-insecurity phase when the individual is not sure whether job loss will or will not occur, the point in time in which the notice of termination is actually given and job loss takes place, and the post-termination unemployment period) each provides its own characteristic significance. In the anticipatory stage, the threat (job loss) has not yet materialized so the problem faced by the individual is one of 'contingency predictability' (Miller, 1981: 204) – the uncertainty as to whether job loss will happen, when it will happen and under what circumstances it will happen. The issues to be evaluated by the individual may include whether, to what extent and how he or she can manage the threat. Can it be prevented? By whom? In what ways? What can be done to minimize the damage? If it cannot be prevented, can it be endured, and if so, how?

On the whole then, while job insecurity and job loss may be highly traumatic and threatening to some, and challenging to others, some key elements involved in one situation may not be generalizable to the other. Even if in real life the two situations sometimes occur in a sequence, in some pivotal respects they differ. This is explored in more detail in the next section.

Distinctions between Job Insecurity and Job Loss

We start our comparison of the properties that distinguish job insecurity from job loss by employing a sociologically oriented perspective. This perspective provides us with important concepts derived from social role theory which illustrate major structural differences between job insecurity and job loss. Three structural distinctions are drawn: the nature of the transition into job insecurity and job loss respectively, the relative social visibility of each state, and the extent to which they are governed by clear role expectations.

The conceptual approach used by role theory is highly congruent with the essentially psychological perspective which is then employed to further elucidate the distinction between job insecurity and job loss. The latter perspective leads us to a discussion of the experiential differences between the two conditions as they relate to the contrasting environmental sources of stress and the different nature of the individual-level crisis involved. Finally, we discuss job insecurity and job loss in terms of the differences between a perceptual phenomenon and an objective state of affairs and consider the implications of this in the light of the evidence from stress research generally.

Role Transition
Originally, the concept of role transition referred to the general phenomenon of moving in and out of roles in a social system (Cottrell, 1942). That the change in employment status involved in job loss falls within the framework of such conceptualization is almost self-evident. Indeed, occupational and professional role gain and role loss are among the major fields of interest in the literature addressing role transitions in the work area (Brett, 1984; Louis, 1980a, 1980b; Nicholson, 1984; Nicholson and West, 1988; Van Maanen and Schein, 1979).

On the other hand, the identification of the passage from relative job security into perceived job insecurity as a role transition is, on the face of it, more problematic. Louis' broader definition of role transition is very useful in this respect. She defines role transition as 'The period during which an individual is either changing roles (taking on a different objective role), or changing orientation to a role already held (altering a subjective state)' (1980a: 330). Recalling our earlier characterization of job insecurity as a transformation of beliefs regarding the present employment relationship, it would appear that this distinction between the change in roles, and the change in orientation to an existing role, applies directly to the analytic distinction between job loss and job insecurity. Thus, in applying Louis' (1980a) terminology, job loss is an *interrole transition*. The central feature of such a transition is the actual crossing of organizational boundaries.

In contrast, the passage from relative job security into job insecurity does not entail any boundary crossing. In Louis' (1980a) typology, it constitutes an *intrarole transition,* namely, a shift in the individual's relation or internal orientation to the presently held employment role. This shift is engendered by changes in individual assumptions concerning oneself, the surrounding internal and external organizational environment, or the relation between self and environment. The intrarole transition into job insecurity, in short, occurs within the familiar employment setting. It is the way in which the individual relates to this setting that is disrupted and can be dramatically altered.

The analytic difference between inter- and intrarole transitions does not in itself imply a qualitative difference between job loss and job insecurity. Certainly, there is no direct indication in Louis' (1980a) typology that the two transition types are different in the kind of adaptational challenges posed. However, in another paper (Sokol and Louis, 1984), the conditions which are likely to help the person in transition are indicated. The emphasis there is on the resources available in the social system in which role transition takes place. In the following paragraphs we suggest that one of the main difficulties in job insecurity is that, unlike job loss, it lacks social visibility and role clarity. Therefore, by implication, it is relatively deprived of such resources. In contrast, the interrole transition involved in job loss, is much more likely to be buffered by some institutional and organizational supports (Schlossberg and Liebowitz, 1980).

Visibility

A phenomenon becomes social when it becomes visible to others, and/ or when this leads to the modification of interaction patterns between the focal person or group and other people or institutions. Of vital importance to our discussion is the fact that job insecurity, in sharp contrast to job loss, is not a socially visible event. Viewed as it is here as a transformation of beliefs, it can more appropriately be described as an *internal event*. To begin with, there is no separation from the physical and social setting of the workplace. For the time being, there is no change in income, work-related activities and relationships or productive social participation. Job insecurity does not involve stigma and deprivation other than possibly in an anticipatory (hence socially invisible) form. While at least according to the supporters of 'deprivation theory' (first and foremost, Jahoda, 1982) the emphasis in job loss and subsequent unemployment is on what is 'taken away', this is not the case in job insecurity. Nothing tangible in the sociological sense (that is, occupational identity, membership and so on) is taken away from or denied to employees who experience job insecurity. In the formal sense, nothing undermines their identities as employed people. Whatever the repercussions of job insecurity, they tend to develop subtly, and their ramifications, if any, proliferate through the role system and beyond without necessarily manifesting themselves in observable formal role changes.

In job loss, on the other hand, there is no lack in social visibility. More often than not it unfolds in an atmosphere of stigma and relative deprivation (Jahoda, 1982; Kelvin and Jarrett, 1985). This is so even though the specific impact on individuals may vary as a function of differences in economic, social and personality attributes. With some exceptions, workers who have lost their jobs and become unemployed are often recognized as unwilling victims who, usually through no fault of their own, have had the instrumental, social and psychological role functions of employment, productivity and income-gaining denied to them. To say that social visibility by itself offers job losers an advantage over insecure employees would of course be a gross overstatement.

Clearly, the direct loss of the instrumental role functions of work, productivity and income-gaining is likely to have repercussions at least as serious as job insecurity throughout the entire role system as well as consequences (economic, social and psychological) for the total person. Despite occasional exceptions, the loss of a job which is critical for subsistence and essential for a sense of productive social participation is likely to induce great stresses. Job loss hits certain role activities immediately and extends to further diverse role relationships, such as the family (for example, Fagin and Little, 1984; Fried, 1979; Hartley, 1987).

To date, there is little evidence on the extent to which job insecurity engenders a similar invasion process into other role relationships. Yet this, and the general absence of social visibility, cannot be taken to imply a negation of a potential for psychological stress in job insecurity. There are even reasons for suggesting it could be greater. As suggested by the stress literature, 'event uncertainty' per se (Lazarus and Folkman, 1984), that is, the subjective probability of an event's occurrence may be an even greater source of anxiety and tension than the event itself. This notion is supported by research evidence from other contexts cited by Lazarus and Folkman (1984), for example, Cohen and Lazarus, 1979; Epstein and Roupenian, 1970; Moos and Tsu, 1977. In our case *event uncertainty* relates to uncertainty about job retention. So job insecurity could well be a potentially equal or even greater source of anxiety than the event (job loss).

Even more importantly, job insecurity, particularly when it is relatively widespread, may be socially constructed. Gossip and rumours within the workplace about who is likely to be retained and who stands a greater chance of being discharged may even deepen the experienced anxiety. The situation is imbued with ambiguity, one aspect of which is the lack of reference groups with which the job insecure can identify themselves and their position in the organization. Job loss, on the other hand, is often experienced by the individual in common with many other people and there may be a greater sense of a shared fate. As such, it provides a sounder basis for identification, however ambivalent, with a clearly defined and institutionalized reference group (others who have lost their jobs). This identification can be of considerable importance. As Brim (1980) points out, it tells the incumbent that he or she is likely to have a support system to help buffer the change. Also, it provides vital sources of role expectations. As will be argued below, it is precisely the certainty of being socially recognized as members of a deprived group that provides the job losers in sharp contrast to the job insecure with, at least, a potential promise of compensatory institutional support.

Role Expectations and Role Clarity

One proposition of role theory contends that the clarity with which roles are defined positively influences the ability to cope with the adaptational challenges involved in the transition from one role to another (Burr, 1972). Role clarity was originally defined by Cottrell (1942: 618) as a set of 'explicit definitions of the reciprocal behaviour expected', as opposed presumably to a set of ambiguous or vague definitions. From this point of view, the normatively governed inter-role transition involved in job loss appears to have some clear advantage over the intrarole transition into job insecurity. Thus, as a result of the social recognition of their status, and from an objective macro-level social vantage point, unemployment following job loss assumes the attributes of a social role that is clearly delineated by certain expectations and responsibilities. Where the normative consensus accepts that those subjected to job loss are not at fault for their condition, the most obvious 'privilege' accorded them is to be exempted from normal 'work'-related activities. For many this may spell disaster. For some, however, as shown in certain studies cited by Hartley and Fryer (1984), this exemption may be associated with a sense of relief from oppressive and dissatisfying features of work – at least until the ramifications of long-term unemployment are confronted more fully. In any case, as long as the job losers are not provided with adequate opportunities to resume the normal obligations of gainfully employed individuals, they are usually entitled to claim assistance and support in the form of welfare benefits, retraining and some help in job search. Although the 'unemployed' role is very much a limited role in comparison with the 'employed' role, and a decrement in activity can usually be taken for granted, the removed tasks or routines are partly replaced by others. The incumbent registers periodically, draws monetary allowances, and is expected to apply for jobs. The job losers, in sum, experience some role clarity and can choose to conform to a socially established role pattern.

The contrast with job insecurity is striking. There is no normative conception that defines it as a social role. It is not institutionalized and no special privileges or responsibilities are associated with it. Without formal notification of imminent displacement, there is no legitimate excuse from the fulfilment of all normal 'employed' role obligations, and the individual who experiences job insecurity is expected to continue his or her job in the organization as effectively as before. The role model he or she is expected to follow is that of the securely employed although some job shrinkage (or expansion) may occur as a result of organizational retrenchment. Ostensibly at least, the major structural dimensions of the employment role as defined by Jahoda (1982) and Warr (1987) remain undisturbed. These include, in addition to regular financial compensation, access to on-the-job social networks, involvement in collective activity, a fixed time structure, and a social identity conferred by the employment context. We do not address ourselves here to the debate over the existence of these dimensions or the quality of experience within each. They need not necessarily be enjoyed or appreciated (Fryer, 1986). Our argument is that as job insecurity sets in, these very dimensions may fail to provide the same role-confirming function as before. At the very core of the job insecurity experience is a growing discrepancy between these dimensions and the employee's interpretations of events and trends. At a minimum, they can no longer be relied upon as effective guides in a subjectively amended state. Thus, while actual job loss and the subsequent transition into the unemployment role provide certain role prescriptions to replace the structural void, the job insecurity experience gives the employee few clues as to role contingencies. No explicit behavioural expectations are available.

The insecure individual may, of course, attempt to compensate for these structural imbalances by trying to develop activities characteristically associated with the unemployment role, for instance by looking for alternative employment. To some extent, the job-insecurity phase may even provide some 'anticipatory socialization' (Merton, 1968: 316) by facilitating an emotional setting for incipient learning and rehearsal of the norms of the 'unemployed' role, before entering that situation. However, if behaviours which are normatively acceptable in unemployment interfere with the obligations of the 'employed' role, the person risks intrarole conflict.

Differences in Environmental Context

On the face of it, some (but not necessarily all) reasons for anxiety during job insecurity and job loss are very much the same. The future becomes less predictable and one is suddenly faced with questions which may not have arisen in continuing and secure employment. What does the future hold? Should I look for a new job or avoid possible failures by not trying? What am I good at? Should I move to another part of the country? How can I reduce my financial commitments? What sources of help can I rely on? In short, both job insecurity and job loss are presumably accompanied by a sense of instability and markedly reduced personal control. Closer examination, however, reveals that both conditions differ qualitatively from each other in some of the aetiological factors with which they are associated.

Job insecurity may be disruptive without necessarily leading to tangible loss (here we treat security as intangible). It is not an event in the sense of necessarily having a clear and discrete temporal onset or termination. On the contrary, it may grow insidiously and become a relatively fixed and ongoing daily experience. Job loss, on the other hand, is usually a relatively eruptive process. By definition, being fired or laid off entails tangible loss, however temporary the loss may be. If we think of crisis as having 'transmission' and 'source' levels of analysis (Hovland and Janis, 1959), then this distinction between job insecurity as an ongoing more chronic phenomenon and job loss as a cataclysmic, sudden and single event may be placed in clearer perspective. Thus, in job insecurity the key issue is 'transmission' of beliefs – will a series of events or agents be perceived as threatening? And how are they apprehended and comprehended, that is, 'Will it happen to me?' Conversely, in job loss the emphasis is on the 'source' – the event itself that requires adaptation.

In job insecurity, any disruption that takes place occurs in the individual's familiar organizational setting. For the time being at least, while there may be a considerable amount of environmental turbulence, the formal relationship between the individual and the organization remains structurally unaffected. Yet the turbulence may lead the individual to perceive the surrounding setting as undergoing change. The salient feature is the kind of uncertainty labelled by Milliken (1987: 136) as 'environmental uncertainty'. Milliken (1987) specifies three types of environmental uncertainty and they all appear readily applicable to job insecurity: (1) *state uncertainty,* the individual's difficulty of predicting how components of the organizational environment might be changing; (2) *effect uncertainty,* the individual's difficulty in predicting what the impact of environmental events or changes will be on his or her prospects for continued job security; and (3) *response uncertainty,* the individual's difficulty in deciding what responses or options are available to him or her, and what the value or utility of each might be, in terms of achieving or increasing security in employment.

In job loss, on the other hand, the basic source of stress lies in the totality of the structural transformation that takes place once the formal employment relationship is severed. Regardless of its manifest forms, stress as experienced in job loss can be viewed primarily as ^ roduct of (1) *environmental discrepancy,* the degree of difference between the employment and the unemployment situations; (2) *environmental congruence,* the 'fit' between the unemployment situation and the person's available resources, and (3) *environmental quality,* the inherent aspects of the new situation without regard to the previous employment situation.

Clearly, the focus on environmental uncertainty in job insecurity does not preclude the simultaneous existence of some of the stress elements inherent in the job-loss experience. Rather, it suggests that job insecurity is an *intermediate* level of experience between the fully secure and job loss. The crux of the matter is whether the meaning placed on events and processes in the organization and its environment is seen as jeopardizing the psychological contract between the individual and the organization and therefore as requiring him or her to made new adaptations. In other words, according to this perspective the definition of job insecurity as a crisis depends on the nature of the worker's phenomenological experience (Jacobson, 1987).

Job Insecurity as a Perceptual Phenomenon
By implication, the foregoing emphasis on the meanings attached to the phenomenological experience of events in the organization and its environment leads us to the last differentiable component of job insecurity discussed here. Unlike actual dismissal or job loss which is unmistakably revealed by the fact itself, job insecurity is cued by one or more inferential ('under the skin') events which are perceived as threatening indicators. In fact, the very presence of job insecurity depends on the individual's interpretations and evaluations of diverse signals in the employing organization's external and internal environments. Some of these antecedent causes for job insecurity have been discussed in detail in the previous chapter. Thus, while job loss or unemployment for the individual who experiences it is an objective state of affairs, job insecurity is a perceptual phenomenon. It denotes what Kurt Lewin (1951) called the 'psychological environment'. It exists inside the person, as a result of his or her perceptions and cognitions of the external objective environment. It is not open to direct observation; rather it is a construct which is inferred – usually from the verbal report of the person or from the observation of behaviour.

Since perceptions vary as a function of contextual factors (Pfeffer, 1983) and as a function of personal attributes (Downey, Hellriegel and Slocum, 1975; Greenhalgh and Jick, 1983; McCaskey, 1976; Thayer, 1967), we may expect considerable variation in the experience of job insecurity in a given organization. Perceptions of reality are likely to differ from 'objective' reality because of limitations in cognitive reasoning abilities and because of emotional responses to threat (MacCrimmon, 1966; March and Simon, 1958). This means that, in situations where objectively all jobs are equally at risk, employees are likely to differ in the amount of job insecurity experienced.

Job insecurity as a perceptual phenomenon is the result of a process that is conceptually close to a cognitive appraisal process. Through this process, according to Lazarus and Folkman (1984: 52–3), 'the person evaluates the significance of what is happening for his or her well-being'. This in itself does not distinguish it from the job-loss experience, which clearly involves a similar process. However, whereas appraisals following job loss are concerned with harm-loss interpretations, that is, with damage already sustained, job-insecurity-linked appraisals are concerned with *anticipated impacts* – the degree to which the present situation is seen as causing possible future disruption. For Lazarus (see, for example, 1966) this primary appraisal of threat (in our case, the threat to job continuity) is determined by two issues. The first, phrased in terms of the question that the focal person asks himself or herself is 'How much am I in danger from the situation?' The second is 'How much am I in danger from anything I do about the threat?' This is independent of whether the objective situation necessitates such an inner dialogue, says Lazarus (1966). The immediate experience of such cognitive processes is followed by attempts to cope with the perceived threat.

The distinction between the cognitive appraisal process that is linked to anticipated impacts (in the case of job insecurity) and that which is linked to damage already sustained (in the case of job loss) can be assumed to have some differential influence on the *severity* of the impact on the individual. This is because job loss relieves at least one major source of stress, that of event uncertainty – the difficulty of predicting the probability of job loss. Job loss is already a certainty so the individual has to come to terms with the loss and cope with its outcomes. In contrast, what makes the job-insecurity experience potentially highly stressful is the fact that coping may for the time being be inhibited by the event uncertainty. Here it is useful to point out that commentators on the stressful economic life-events literature (for example, Dooley and Catalano, 1980; Dooley et al., 1986) agree that undesirable financial or job-related outcomes are not necessary for adaptation syndromes to occur. The mere anticipation of such an outcome may precipitate measurable stress reactions (see also Kinicki, 1985; Swinburne, 1981). Stress research in other areas as well suggests that cognitive or perceptual conditions are *at least* as relevant in predicting responses as is the objective intensity of the stressor. Thus, as already indicated earlier in this chapter, in a number of studies it was found that dealing with uncertainty about potentially disastrous and poorly predictable future events may, in fact, create even greater psychological stress than the events themselves. Examples to this effect cited by Lazarus and Folkman (1984) included women whose husbands were reported missing in action, in comparison to wives whose husbands were killed, and cancer patients who were unsure for many years whether or not a cure had been effected.

Conceptualizing and Measuring Job Insecurity

In defining job insecurity, three distinctions must be considered: insecurity as an objective or subjective phenomenon, insecurity as a cognitive or affective quality, and insecurity about the continuity of one's job or aspects of one's job.

In the previous section we have already explored subjective versus objective job insecurity. Job insecurity, as we see it, is a subjective phenomenon: cues that are objectively available are perceived as threats to the continuity of one's job. The second dimension – cognitive versus affective – relates to beliefs and feelings about insecurity. On the one hand, job insecurity refers to the cognitive component of the likelihood of losing one's job. 'The less likely it is in my eyes that I will keep my job, the more insecure I am.' On the other hand, insecurity encompasses the affective component of being concerned. 'If I were to lose my job, how much am I concerned about it?' Psychologically speaking, both the affective and cognitive components are important. 'I can only be worried about losing my job if I think that it is likely that I will lose it. But, if I do not care about my job, the likelihood of losing my job is of little concern to me.' Finally, the concept of job insecurity has been used to signify both fear about the continuity of one's job and aspects of one's job. In studies on work-related stress, it usually relates to aspects of one's job (Dijkhuizen, 1980). Although we do not deny that uncertainty about the continuity of specific aspects of one's job can be of interest we will concentrate in this book on threats to the continuity of the total job.

Such a conceptualization is in line with Greenhalgh and Rosenblatt's (1984: 438) definition of job insecurity as 'perceived powerlessness to maintain desired continuity in a threatened job situation'. Critical to feelings of job insecurity is the individual's felt vulnerability originating with his or her perception of threatening signals in the work environment.

Our use of the term 'vulnerability' draws on Lazarus and Folkman's (1984: 50–1) definition of psychological vulnerability as a 'deficiency in resources', but 'only when the deficit refers to something that matters'. 'Vulnerability', they suggest 'can be thought of as *potential threat* that is transformed into *active threat* when that which is valued is actually put in jeopardy'. Lazarus and Folkman's (1984) approach emphasizes the characteristics of the individual, on the one hand, and the nature of the environment on the other. In defining stress, they speak of 'a relationship between the person and the environment that is appraised by the person as taxing or exceeding his or her resources and endangering his or her well-being' (Lazarus and Folkman, 1984: 21). If we accept this interactionist definition of stress, then we can say that the extent to which changes in the work environment lead to a stressful perception of job insecurity hinges on three major factors: (1) the beliefs about what is happening in the environment, that is, the appraisal of the threat posed by change; (2) the resources available to the individual, as perceived by him or her, to counteract the threat; and (3) the perceived seriousness for the individual of the consequences if the threat actually happens (that is, job loss).

Greenhalgh and Rosenblatt's (1984) description can serve as a starting point for our elaboration of the concept of job insecurity. Their approach to job insecurity has two basic dimensions: the severity of the threat to one's job and the extent of one's powerlessness to counteract the threat. According to Greenhalgh and Rosenblatt (1984), these two dimensions have a multiplicative relationship, as the following equation illustrates:

$$\text{felt job insecurity} = \text{perceived severity of the threat}$$
$$\times \text{ perceived powerlessness to resist threat}$$

The multiplicative relationship signifies the assumption that employees only feel insecure about their jobs if they perceive the threat to be severe *and* feel powerless. Employees who either do not care or who feel capable of resisting the threat to their jobs are presumed to feel no job insecurity.

Greenhalgh and Rosenblatt (1984) maintain that the perceived severity of the threat depends upon the importance the individual attaches to potential work-related losses due to job discontinuity, and the subjective probability that these losses will occur. These perceptions combine multiplicatively as well; if either of the dimensions is not very great, then the degree of job insecurity will not be very great:

$$\text{perceived severity} = \text{importance of job features}$$
$$\times \text{ likelihood of losing job features}$$

Jacobson's (1985) 'Job-at-Risk' model retains Greenhalgh and Rosenblatt's (1984) multiplicative conceptualization of the 'felt job insecurity' as a cognitive construct. Following the Lazarus and Folkman (1984) approach, however, Jacobson's model reformulates powerlessness more generally as one component of 'perceived susceptibility'. The model posits that the extent to which felt job insecurity is aroused is determined both by the employees' perceived susceptibility and their perceptions of the consequences of job loss. The differences in the degree of job insecurity felt by given individuals in the same organization will reflect each individual's specific aggregation of susceptibility factors on the one hand and 'perceived severity' on the other. Thus, in line with Greenhalgh and Rosenblatt (1984), the Jacobson (1985) model suggests that the employee who believes that he or she is susceptible will conduct a cognitive calculus involving the subjective importance of each life-situation feature that could be endangered (or improved) as a result of job loss and the subjective probability that it will be endangered (or improved).

A similar modification of the Greenhalgh and Rosenblatt (1984) conceptualization has recently been proposed by Brown-Johnson (1987), who suggests that powerlessness can be incorporated as part of the probability of the loss, since powerlessness to resist the threat makes the loss more likely. If employees perceive that they have more power, their assessment of the probability of job loss will decrease. Thus, according to Brown-Johnson (1987), powerlessness is not conceptually distinct from the perceived probability of job loss. In line with the Jacobson (1985) and the Brown-Johnson (1987) revision of the Greenhalgh and Rosenblatt (1984) conceptualization, perceived probability of job loss can essentially be expressed in expectancy theory terms. This theory analyses an individual's attitude towards a specific event as a function of the probability of the event and the value of the expected outcomes of its occurrence (Feather, 1982; Mitchell, 1982; Porter and Lawler, 1968). In the present case, the perceived probability of job loss would represent an expectancy component, and the severity of job loss a value component.

Global versus Multidimensional Measurement of Job Insecurity
The explicit operationalization of perceived job insecurity is only just evolving in organizational psychology research. Only a handful of studies addressing the measurement of job insecurity exist, and most of these suggest more or less the same type of global variable for measuring it. The most recent review of these studies is provided by Ashford et al. (1989). These authors point out that earlier studies of work and stress have treated job insecurity in an ad-hoc manner as one of a group of potential job-related stressors. They cite the studies by Caplan et al. (1980) and Johnson et al. (1984) as examples of this approach.

Caplan et al. (1980: 45) developed a four-item 'Job Future Ambiguity' scale in their comprehensive study of fourteen stressors in the objective and subjective work environment. Their model categorizes job future ambiguity as a subjective stressor and refers to 'the amount of certainty the person has about his job security and career security in the future'. Their results show high job future ambiguity to be one of several stressors associated with job dissatisfaction.

Johnson et al. (1984) generated a seven-item job-insecurity subscale from an initial pool of positive and negative statements reflecting attitudes one might hold towards work. Sample statements included: 'The thought of getting fired really scares me'; 'I am worried about the possibility of being fired'; and 'I am so worried that I would do almost anything to keep my job.' Respondents were asked to indicate the extent of their agreement on a five-point Likert-type scale. The job-insecurity subscale was then combined with four other subscales to produce a Work Opinion Questionnaire (WOQ). The investigators reported that the job-insecurity subscale scores contribute to the prediction of work performance on entry-level jobs among low-income youth and adults.

Whereas in the forementioned studies job insecurity was measured as only one out of a number of stressors, Greenhalgh (1982, 1985 and

Chapter 8 in this volume) more specifically investigated the relationship between job insecurity and indicators of organizational effectiveness. He supplemented Caplan et al.'s (1980) job future ambiguity scale with a global item ('How certain are you about your job security?'). Following factor-analysis, the global item and two of the original items were retained. The latter two read: 'How certain are you about what your future career picture looks like?'; and 'How certain are you of the opportunities for promotion and advancement which will exist in the next few years?' Here too a five-point Likert-type scale offered response categories ranging from 'very uncertain' to 'very certain' (Greenhalgh, 1985).

Both Greenhalgh and Rosenblatt (1984) and Ashford et al. (1989) have argued in favour of what they called a multidimensional operationalization of job insecurity. Indeed, Ashford et al. (1989) attempted to test the construct validity of a job-insecurity measure based on Greenhalgh and Rosenblatt's (1984) conceptualization. Separate scales were constructed to measure various components: importance of job features, likelihood of losing job features, importance of job loss, likelihood of job loss and perceived powerlessness to resist threat. The multiplicative combination of these components as prescribed by Greenhalgh and Rosenblatt's (1984) model formed Ashford et al.'s (1989) Overall Job Insecurity measure. They compared their overall measure with the global measures of job insecurity developed by Caplan et al. (1980) and Johnson et al. (1984). The three job-insecurity measures were correlated to several outcome variables such as turnover intention, commitment to the organization, trust in the organization, satisfaction, somatic complaints and job performance. On average, the multiplicative job-insecurity measure explained more variance in the dependent variables than did either the global Caplan scale or the Johnson job-insecurity scale.

The importance of Ashford et al.'s (1989) work lies in its rigorous operationalization of the elements of Greenhalgh and Rosenblatt's (1984) job-insecurity model. Although Ashford et al. (1989) warn against inferring too much from a single study, they clearly intend to encourage the use of multidimensional measures of job insecurity.

There are, however, arguments for keeping the elements of the job-insecurity model separate, rather than combining them into a single multidimensional measure of job insecurity. Global measures that restrict themselves to assessing feelings of job insecurity per se, rather than combining them with elements of an explanatory model, have the advantage of keeping open the possibility of posing theoretical and empirical questions about the relationships between elements of the model. Such questions are not pertinent if those elements are combined into one single measurement. In that regard it seems no accident that Ashford et al. (1989) do not discuss the very interesting finding that the powerlessness dimension of their multidimensional scale is the most important component in relation to the antecedents and consequences of job insecurity. Similarly, they do not examine the significant differences in the correlations of individual components. In addition to this theoretical argument, there is the more practical question of parsimony. As Cooper and Richardson (1986) argue, variables of greater elaboration usually explain more variance. It is not very surprising, these writers would certainly argue, to obtain better results with a variable that is comprised of sixty items compared to variables that are construed on only four or seven items. Because it is not always possible, however, to administer such long questionnaires, it is important to know how much of an improvement in terms of explained variance is achieved. In this regard Ashford et al.'s (1989) results are disappointing. It is interesting to see that one of the components of Ashford et al.'s (1989) overall measurement – powerlessness – is almost as good as the overall measurement, although it contains not more than three items.

Consequently, theoretical and practical considerations have to be taken into account in choosing between global and multidimensional measurement strategies. Both considerations convinced us to prefer the use of global measures; in other circumstances with different research questions a multidimensional approach might be more appropriate.

The Approach to the Measurement of Job Insecurity in this Book
We conclude this chapter by briefly describing the operationalization of job insecurity in the Israeli, Dutch and British studies in this book.

The measure used in the *Israeli* study consisted of three items. The first was a subjective evaluation of the likelihood that the respondent would lose his or her job: 'To what extent in your opinion are you likely to lose your job in the near future?' The second and third items gauged the respondents' present feeling of employment security: 'To what extent, in your judgement, are you likely to be employed in your present job three months from now?' and 'To what extent in your opinion are you presently safe from dismissal in your current place of employment?' Each item had a five-point Likert-type response scale. The measure yielded an inter-item reliability of .82.

The *Dutch* study also employed a three-item measure, though with different wording. The first item, subjective likelihood, read: 'At the present time, does it seem likely that you will lose your job?' This item had a three-point Likert-type response scale. The second item was on concern about job insecurity: 'To what extent are you worried at the present time about your continued employment in your current job?' (four-point response scale). The third item addressed the individual's satisfaction with job security: 'To what extent are you satisfied with the security provided in your present job?' (five-point response scale). The total job-insecurity score was arrived at by standardizing and adding the scores on individual items. The inter-item reliability of this measure was .77 (Cronbach's alpha).

The *UK* study also employed a three-item measure. The first two items involved the subjective likelihood of losing one's job: 'How secure do you feel in your present employment?' and 'I feel sure that my job will continue here for a long time.' The third item addressed concern about job loss: 'I have no worries about the future of my job.' Here too each item offered a five-point Likert-type response scale. The inter-item reliability was .75 (Cronbach's alpha).

There is one important difference between the Israeli measure on the one hand and the Dutch and UK measures on the other. Whereas the Israeli measure focuses on the cognitive 'likelihood of job loss' dimension, the two other measures encompass in addition the affective 'concern over job loss' dimension. This difference does not result from diverging opinions among ourselves, but from the different approaches taken before the beginning of our collaboration.

Our discussion so far has shown that job insecurity can be operationalized in different ways. At this early stage of the research field there is no need to reach a consensus. Different approaches can be creative because they focus attention on different facets and stimulate further investigation. To accept closure of discussion too early could unnecessarily restrict the field.

Summary

In this chapter the topic of job insecurity as a particular experience in the work situation was introduced. We began with a systematic analysis of the ways in which job insecurity differs conceptually from job loss. Using terminology borrowed from role theory, job insecurity was presented as an intrarole transition, and job loss as an interrole transition. The differences that emanate from that distinction relate primarily to some structural properties of the role dynamics involved. We suggested that job insecurity can be viewed as an intermediate category which, unlike job loss, has little social visibility and no specific normatively governed role prescriptions attached to it. It is an *internal event* reflecting a *transformation of beliefs* about what is happening in the organization and its environment. Job insecurity may acquire its presence insidiously and become a relatively fixed and ongoing daily experience.

One of the salient features of job insecurity is its association with various forms of event and environmental uncertainty. As such, it has the potential of becoming even more stressful than job loss insofar as the coping process may be inhibited by the uncertainty of the event.

In job loss, on the other hand, this major source of stress, event uncertainty, has been removed, and the individual is faced with the need to come to terms with the loss and to face the outcomes of the event itself. The sources of stress and tensions as job insecurity is encountered are not necessarily the same as when the job is lost.

Above all, the distinction between job insecurity and job loss is a distinction between a purely subjective experience and an objective reality which may, of course, also have some subjective substance. This means that in organizations where objectively all jobs are equally at risk, employees are likely to differ in the amount of job insecurity experienced. Precisely therein lies the special challenge of understanding job insecurity better, for it is so easy to miss the power of subjective and latent realities which have little visible substance.

Drawing on these distinctions between job insecurity and job loss we concluded that a stressful perception of job insecurity hinges on three major factors: (1) the cognitive appraisal of the threat posed by the beliefs about what is changing in the work environment; (2) the resources available to the individual; and (3) the perceived seriousness of the threat to job continuity should it materialize. In line with earlier elaborations in the literature, job insecurity is thus conceptualized as an interaction between the perceived probability and the perceived severity of losing one's job, where 'severity' is a function of the subjective importance of each home and work feature that could be endangered by job loss, and the subjective probability of it being endangered.

Following this, we reviewed earlier attempts to operationalize the job-insecurity construct. We took issue with the only empirical attempt thus far, based on the foregoing conceptualization, to assess the utility of a multidimensional measure and to determine its construct validity. On the basis of theoretical and methodological considerations, we suggested that it is preferable to keep the constituent elements of this conceptualization apart and to study their respective impact separately. This allows for a more parsimonious global definition of job insecurity as 'a concern about the future of one's job'. This definition was adopted as a basis for the specific measures used in the Israeli, Dutch and British studies.

3

Employees and Job Insecurity

Bert Klandermans, Tinka van Vuuren and Dan Jacobson

Social psychologists have repeatedly shown that different individuals will arrive at different definitions of situations that objectively appear to be identical. Thus, we may expect employees who are exposed to the same threatening signals to experience differing degrees of job insecurity. On the other hand, people do not live in isolation, so an individual's definition of the situation can be at least to some extent socially constructed, that is, developed through interaction with other people. Consequently, employees may share their feelings of job insecurity with others who are in the same situation. The conditions that make employees diverge or converge in their definition of the situation are far from clear, but we feel it is worthwhile to try to acquire more insight into these conditions.

In our previous chapter we argued that job insecurity can be a stressful experience for some people. It seems plausible, then, that the greater the stress produced by job insecurity, the more adverse its impact. As the literature on stress suggests, reactions to stressful situations vary considerably. In the case of job insecurity, we may expect a similar variety of reactions, as our discussion in Chapter 2 indicates. We will now expand this into a framework for the analysis of individual responses to job insecurity.

Both this theoretical chapter and the following empirical chapters attempt to answer three central questions that any analysis at the individual level must consider: What makes employees feel insecure about their jobs? How do employees respond to feelings of job insecurity? What explains the differences in responses to job insecurity? Basically, these questions concern the predictors and the consequences of job insecurity. We can identify predictors and consequences at three levels: at the individual, the organizational, and the industrial-relations levels. This theoretical chapter and the next two empirical chapters will focus on the individual level; the organizational and the industrial-relations levels are the subject of later chapters.

Fears and Hopes: Predicting Job Insecurity

Job insecurity is a mixture of fears and hopes about the future of one's job. What events, circumstances, signals, characteristics make employees feel insecure? Since we conceive of job insecurity as a function of the perceived probability and the perceived severity of losing one's job, every factor, condition or circumstance that influences the perceived probability, the perceived severity or both may increase job insecurity. The more likely it is that a person will lose his or her job, and/or the more severe the consequences of the loss, the stronger his or her feelings of job insecurity will be. For the scientist and the practitioner who want to get some grip on the phenomenon of job insecurity it is important to know what factors predict the appearance of job insecurity. If we know some of the predictors we may be able to control better those situations in which jobs appear to be threatened, and to prevent the development of unrealistic pessimism or optimism among employees.

The Probability of Losing One's Job
What cues in the environment, what characteristics of the situation or of the person can help us predict whether an individual employee will fear losing his or her job? If job insecurity is a perceptual phenomenon, will feelings of insecurity depend completely on the individual? Will objective circumstances leave their mark on an individual's perception, or is job insecurity a completely subjective response to the situation? In this section, we hypothesize that the perceived probability of losing one's job is the result both of characteristics of the individual employee and of the situation.

In their discussion of an employee's powerlessness to counteract an experienced threat to his or her job, Greenhalgh and Rosenblatt (1984) describe the forms powerlessness can take as (1) lack of protection, (2) the culture of the organization, and (3) the type of standard procedures for dismissing employees. Since powerlessness is an aspect of the perceived probability, these same factors can be seen as influencing the perceived probability of losing one's job or aspects of one's job. Lack of protection refers to the absence or weakness of such elements as unions, seniority systems and employment contracts that can serve to protect individual employees. The culture of the organization may increase an employee's sense of powerlessness when there are no distinct norms regulating employment relations, when employees have little influence on decision-making, or when the evaluations and decisions of superiors are seen as arbitrary. Finally, standard procedures for dismissal can make employees more or less vulnerable (see Chapter 9 for a more fully elaborated treatment of this subject).

The degree of felt personal susceptibility to job loss in the light of threatening cues can be expected to vary widely among individuals in a given organization. Jacobson (1985) introduced the notion of perceived personal susceptibility to job loss to explain this variation. Numerous factors may be responsible for individual differences in perceived susceptibility. For example, some employees may perceive themselves to be susceptible to job loss because they feel that their performance does not meet the standard. The degree of protection employees expect from their relative seniority, or from their union, may also colour their perceptions. Still others may believe that they are indispensable to the organization.

Whereas Greenhalgh and Rosenblatt (1984) discuss organizational characteristics that may intensify job insecurity, and Jacobson (1985) points to perceptions as modifiers of job insecurity, Ashford et al. (1987), in distinguishing between role ambiguity, role conflict, locus of control and anticipated organizational change refer to organizational characteristics and personality characteristics as predictors of job insecurity. In their study, only locus of control turned out to be a significant determinant of feelings of job insecurity.

Although none of the three publications alone draws a complete picture, together they establish at least the outlines of a systematic treatment of factors influencing the perceived probability of losing one's job. Such factors exist on three different levels: (1) characteristics of the organizational climate, or more generally, the industrial-relations climate within the organization; (2) individual characteristics of the employees and/or of their position in the company; and (3) the employee's personality characteristics. At each of these levels potentially influential factors appear. The organizational or industrial-relations climate is determined by such factors as individual or collective contract regulations, standard procedures for dismissal, trust in management, and the perceived strength of the unions or workers' councils. Individual or positional characteristics involve one's health, seniority, position in the company, education, ethnic background, relationship to supervisor or management, and work experience. Examples of personality characteristics that may influence the perceived probability of job loss are locus of control (Ashford et al., 1987; Büssing and Jochum, 1986) or individual predispositions to optimism or pessimism about one's future situation. Although additional factors may play a role in each of the three categories, for the moment it suffices to specify these categories and to provide some illustrative examples.

The Severity of Losing One's Job
Job loss is usually seen negatively, that is, in terms of the degree to which it is likely to harm the individual's chances of achieving, obtaining or maintaining important values, resources or objectives. But in

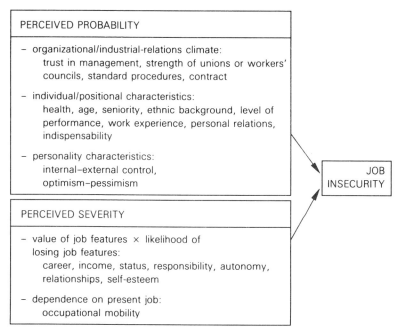

Figure 3.1 *Predictors of job insecurity*

some cases it may also be viewed positively, as a challenge, a relief or an opportunity for growth or development. This latter view might become more prevalent if job loss could be seen as a 'breaking away' from a negative reality involving boredom, fatigue, interpersonal conflict and psychological or physical strain. Admittedly, however, the value attached to paid employment in contemporary Western society suggests that, at present, most people usually feel that the costs of job loss outweigh the gains (Warr, 1984).

Once employees believe they are susceptible to job loss, they become aware of the subjective valence of the job features that are endangered. Job features an individual may fear losing include career progression, income, status, self-esteem, interpersonal relations, responsibility and autonomy. The more an individual values these features, the more severe will be the effects of their loss. The prospect of such losses is all the more threatening if individuals are very dependent on their current jobs (Greenhalgh and Rosenblatt, 1984). Individuals who feel that their occupation makes it easy for them to find another job presumably tend to be less concerned with security than are employees who feel they have few alternatives.

In sum, the more an employee values his present job, and the more he depends on it for acquiring valued features of that job, the greater the perceived severity of job loss.

Summary
Job insecurity is presumed to be a function of the perceived probability and the perceived severity of losing one's job (see Figure 3.1). Factors influencing either of these two dimensions may determine an individual's feelings of job insecurity. The perceived probability of losing one's job is influenced by characteristics of the organizational or industrial-relations climate, characteristics of the person or his or her position in the organization, and by personality characteristics. The perceived severity of losing one's job is affected by the value of various job features, the likelihood of losing these features, and the employee's dependence on his or her present job for these particular features.

Coping and Well-being: Responses to Feelings of Job Insecurity

Job insecurity can be a stressful experience that unquestionably influences an individual's well-being and job performance. Job insecurity can be one of the more important stressors in employment situations (Kahn et al., 1964; Dijkhuizen, 1980), and Greenhalgh and Rosenblatt (1984) even go so far as to draw a parallel between uncertainty about the continuation of one's job and the anticipatory stage in the process of bereavement. Until now, few researchers have conducted systematic studies of reactions to job insecurity per se. The few studies that are available restrict themselves primarily to work-related responses, such as commitment to the organization, turnover intent, job performance, job satisfaction (Ashford et al., 1987), and perceived influence in the organization (Büssing and Jochum, 1986). Greenhalgh and Rosenblatt (1984) also concentrate on employees' performance in situations of job insecurity. They discuss several responses to feelings of job insecurity: reduced effort, propensity to leave and resistance to change. As Greenhalgh and Sutton argue in Chapter 8, on the organizational level these reactions can easily start a vicious circle by inducing a further decline of the company, which in turn jeopardizes more jobs. Job insecurity, however, influences a much broader range of feelings, attitudes and behaviours than those related simply to performance. Jacobson (1985) was the first to bring the study of reactions to job insecurity fully within the coping-with-stress framework. Although job insecurity has always been recognized as a work-related stressor, and although literature on coping with stress abounds, until Jacobson developed his 'Job-at-Risk' model, no one had provided a separate analysis of coping with job insecurity. Relying on Jacobson's adaptation of the stress framework, we will discuss different strategies for coping with job insecurity and then try to explain why different individuals adopt different coping strategies.

Stressors, Strains and Coping with Stress
Stress, stressors, strains and coping with stress are the core concepts
of the stress framework. But before we set out to apply this framework
to the context of job insecurity, we must define the basic elements of
the phenomenon we are describing: What is stress? What are stressors
and strains, and what are the different ways of coping with stress?
Stress has been defined in three different ways: as a stimulus, as a
response and as an interaction (Dijkhuizen, 1980; Kleber, 1982; Buunk
and de Wolff, 1989). The definition of stress we use here is the
interactionist definition proposed by Lazarus and Folkman (1984) and
French et al. (1982). These authors define stress as a situation in which
the demands of the environment exceed the individual's capacity. The
term 'stressor' refers to stress as a stimulus: any event that an
individual may experience as threatening, uncomfortable or unpleasant.
All kinds of events or situations can function as stressors, for example,
interpersonal conflict, or illness or, indeed, job insecurity. Under these
circumstances individuals may experience tension. And, according to
those who define stress as a response, this tension produced by a
perceived threat is what constitutes stress. Psychosomatic symptoms of
the tension occasioned by stressors (nervousness, sleeplessness,
headaches and so on) are called 'strains'. The term 'coping with stress'
refers to the various ways different individuals deal with stress. Coping
is the outcome of the cognitive appraisal process (see Chapter 2) in
which the individual evaluates both the threat and the alternatives for
dealing with that threat. In Lazarus's view (1966), the two major functions of coping are: first, to change the situation for the better if one
can, either by changing one's own action or by changing the threatening environment; and, second, to eliminate emotional stress. Consequently, we can distinguish two forms of coping behaviour (Lazarus
and Folkman, 1984): problem-directed behaviour – intended to remove
the unpleasant event or to mitigate its influence; and emotion-directed
behaviour – intended to alleviate the distressing feelings caused by the
unpleasant event, for instance, by way of relaxation, alcohol or defence
mechanisms such as denial. These distinctions parallel very closely the
two general ways of coping described by Winnubst (1980): 'active
response' and 'avoidance'. Through active responses individuals seek
to change their situation; whereas avoidance, on the other hand,
indicates a withdrawal from the situation.

Coping with Job Insecurity
What are the different ways of coping with job insecurity? Can we
adopt the distinction between avoidance and active response, or do we
need additional modes to map the way people deal with threats to the
future continuation of their jobs? To answer this question we return to
the literature on job insecurity. Greenhalgh and Rosenblatt (1984) do

not specifically discuss different strategies for dealing with job insecurity. The consequences they discuss are relevant in an organizational context but do not cover the full range of individual responses to job insecurity. Jacobson (1985) expanded Greenhalgh and Rosenblatt's theory of job insecurity by adding the concept of coping with job insecurity and by categorizing different coping strategies. It is to these coping strategies that we now turn.

The four coping patterns Jacobson (1985) distinguished are based on the work of Janis and Mann (1977) on coping with threatening circumstances such as earthquakes. The four patterns are: unconflicted inertia, unconflicted change, defensive avoidance and vigilance.

Unconflicted inertia can be expected when the threat of job loss evokes no reaction, that is, when individuals do not feel any job insecurity either because they do not believe they personally are susceptible to job loss or because they do not believe that job loss is likely to have an adverse effect on their lives, or both. Under such circumstances, it is assumed, individuals will continue their work behaviour as before, since they are not aware of any serious threat to their job security.

When the individual scores relatively highly on both dimensions of felt job insecurity (that is, perceived probability and perceived severity), one of the three remaining coping patterns may emerge, depending on the availability of cues and the individual's assessment of the benefits of action weighed against the barriers or costs.

When the costs of a given course of action are perceived as relatively low and the barriers are perceived as relatively insignificant, and when the belief in the potential benefits and effectiveness of the action is relatively high, we suggest that the probable response will be *unconflicted change*. This response is an adaptive, action-oriented coping pattern that can be aimed at the self or the organizational environment, or both. Actions focused on the self may consist, for example, of increasing one's investment in job performance to prove one's indispensability to the organization, thereby (one hopes) averting susceptibility to job loss. Such action may also involve an energetic and deliberate search for alternative job opportunities, as well as re-education and retraining, use of counselling services, reserving money for financial support, and, ultimately, voluntary departure. An unconflicted change pattern directed at the environment may additionally involve engaging in intra- or extra-organizational political action to reduce the threat of job loss or to minimize the severity of the consequences of dismissals.

When felt job insecurity is great but the barriers to action are also seen as formidable, when the feasibility and efficaciousness of protective action are viewed pessimistically, or when options for such action are perceived as unavailable altogether, we can expect the *defensive avoidance* coping pattern to prevail. Defensive avoidance involves an acceptance, albeit a negative one, of the reality of the threat. In this situation, the individuals' fears are highly aroused and the threatened job loss increases the vigour of their initial response. But the employees are equally, or even more strongly, deterred and repelled by the perceived costs and barriers associated with action. They will therefore tend to remove themselves psychologically from the situation by avoiding cues to action. Subsequently, they may disengage themselves and develop a general lack of interest or fatalistic attitudes, perhaps distracting themselves with other irrelevant activities. According to Janis and Mann (1977), we may also see defensive avoidance as helpless dependency on others to make decisions, rationalization ('others know better'), and a tendency to shift to others all responsibility for finding a solution – the 'others' usually being management, the workers' council, the union or the government.

Finally, when action in the direction of 'change' is perceived as a potentially effective and beneficial means of handling insecurity but the costs are viewed as equally high, we can expect the *vigilance* coping pattern to predominate over defensive avoidance tendencies. Although at this stage employees consciously refrain from committing themselves to a given course of action, such as applying for a new job, they search for relevant cues (for example, they regularly scan the newspaper listings of job openings) needed to make a sound coping decision or to reappraise the threat to their job insecurity. This activity can also include scanning the cues within the organization (to check on rumours, for example), going to meetings for information, discussing the organization's performance, and so on.

Jacobson's (1985) treatment of coping with job insecurity raises several unresolved questions. First, it would be difficult, if not impossible, to distinguish his unconflicted inertia pattern from the psychological state of individuals in situations where no job insecurity exists. In fact, we may wonder whether unconflicted inertia is a coping pattern at all, because there is no need to engage in coping when one feels no insecurity about one's job.

Second, it would be empirically difficult to separate vigilance from unconflicted change. Referring to perceived costs and benefits is not of much help, for how can we determine that the barriers and costs of acting are equally as high as the perceived benefits? Further, quite often vigilance (as described) will be the first step in an unconflicted change strategy. Consequently, we believe that, at least empirically, the four coping strategies can be reduced to two basic forms of coping: *avoidance* and *active response*. 'Avoidance' as we use it is identical to Jacobson's definition of the term; 'active response' combines Jacobson's unconflicted change and vigilance.

Third, although Jacobson correctly mentions the possibility of becoming engaged in intra- and extra-organizational political actions (for example, by joining a union, actively participating in union activities or participating in industrial action), his study, like the stress literature in general, does not provide a systematic treatment of collective responses to stress. We must note that studies of collective reactions to stress are rare, even in the case of situations where collective reactions are not unusual, for example, among employees with common interests. Sometimes employees will protest collectively against measures taken by their company's management which jeopardize their jobs or relevant aspects of their jobs. But biased towards the individual as the stress research is, few studies have, until now, paid attention to shared strains and collective responses in reaction to stress. In the stress researcher's defence, one can argue that collective reactions to common stressors are rare indeed, not because common stressors rarely occur, but simply because so often the conditions for collective actions are not met. Collective responses to job insecurity, we will argue, are not just another type of active response to a stressful situation; they are qualitatively different from individual active responses. We will develop this argument later in this chapter. At this point, however, a look at Hirschman's (1970) discussion of exit and voice as responses to decline in organizations may help to clarify our view.

According to Hirschman (1970), when an organization is declining, essentially two kinds of responses are open to employees: exit and voice, that is leaving the company or expressing criticism aimed at improving the situation. Exit from the organization is usually an individual reaction; verbal protest is more likely to be a collective response (Tajfel, 1975). Hirschman (1970) notes that criticism has a greater impact on the organization if it is offered by those who can afford to leave instead of by employees who have no other job options. The better qualified, those who can get another job most easily, are those whose voices are most effective. Hirschman (1970) argues that the organization will be more amenable to their demands when their criticism is accompanied by threats of quitting. On the other hand, the easier it is for an individual to find another job, the more likely it is that he or she will in fact quit. In other words those best equipped to speak up may be those who are the first to quit. The following example from a Dutch company illustrates that employees are well aware of this mechanism and are even capable of developing a strategy to counteract it.

In this company, which was facing decline, the trade union lay officials were confronted with the choice of either looking for other jobs or initiating action to retain jobs in the factory. It was evident that it made no sense to initiate action unless many lay officials could be counted on to support the action for an extended period of time, so the union representatives agreed that they would continue in their jobs for one more year. Should there be no improvement in the situation after that time, everyone would be free to move on. The lay officials honoured their agreement and did not apply for jobs in other firms until a year later. Of course, such demonstrations of cohesiveness do not always occur. In Chapter 7 an example of the opposite phenomenon – sectionalism among the workforce in response to job insecurity – will be described.

To summarize, we can expand our framework for the study of job insecurity by including avoidance and active responses as the two basic forms of coping with job insecurity, and by distinguishing between individual and collective action as two qualitatively different forms of responding actively to job insecurity.

Before we turn to our last question – what makes employees adopt a specific coping strategy? – we must consider one more topic that both Greenhalgh and Rosenblatt (1984) and Jacobson (1985) mention but do not pursue: the strains related to job insecurity.

Psychosocial Well-being

Stressful situations often result in the 'reduced psychological well-being' of those affected. Psychological well-being involves a whole complex of mutually interrelated affective, cognitive and behavioural processes. Impaired psychological well-being is characterized by such phenomena as anxiety, depression, a sense of uselessness, lack of self-confidence and dissatisfaction with oneself and with one's environment (Warr, 1984, 1987). Our question is whether job insecurity, like other stressors, reduces employees' psychological well-being. Do people feel depressed, anxious, useless and dissatisfied with their work environment if they fear losing their jobs? The available evidence suggests an affirmative answer.

The literature on work-related stress consistently reveals negative correlations between stress and satisfaction (Brief and Atieh, 1987; Caplan et al., 1975; Dijkhuizen, 1980); the more stress individuals experience, the less satisfied they are with their work environment (their job, their work organization, and so on). Consequently, we may hypothesize that job insecurity as a work-related stressor will similarly have negative correlations with job satisfaction and satisfaction with the organization.

Research on job insecurity and psychological health, although limited, has produced divergent results. Dijkhuizen (1980) found that a feeling of job insecurity leads to anxiety, depression and irritation. Kuhnert (1987) also concluded that job insecurity and psychological health are intimately connected. The higher the employees estimated their chances of retaining their jobs – that is, the more secure they felt – the greater their psychological health. Catalano et al. (1986)

demonstrated that individuals who feel insecure about their jobs are more likely both to consider seeking help and to actually seek help for psychological problems than are people who feel no job insecurity.

In contrast, Depolo and Sarchielli (1985) and Büssing and Jochum (1986) found no relation between job insecurity and psychological health. Depolo and Sarchielli inquired into the well-being of employees confronted with a decline in their factory. Contrary to expectation, their results showed that employees who continued to be employed at the plant felt no better than employees who were temporarily laid off because of reductions in working hours. Depolo and Sarchielli explained this finding by arguing that those who were temporarily unemployed experienced the same degree of insecurity as those who continued on the job. Büssing and Jochum (1986) examined reactions to job insecurity by using a quasi-experimental research design. They compared the reactions of employees in a metal processing firm where jobs were not endangered with employees' reactions in an enterprise where they were. Because psychosomatic complaints were equally frequent in both groups, Büssing and Jochum concluded that job insecurity does not seem to lead to a reduction in psychological well-being.

Neither the German nor the Italian researchers, however, sought to determine the degree to which employees did or did not experience job insecurity. Instead, they studied job insecurity by comparing two objective conditions. Depolo and Sarchielli compared the well-being of those who remained at home because of reduced working hours with the well-being of those who continued to work, while Büssing and Jochum compared employees in a firm where employment was assured with those in a firm where jobs were at risk. But if job insecurity is considered to be a subjective experience based on employee perceptions, then similar or identical objective situations may produce differing degrees of insecurity in different individuals. And the perceived probability or the perceived severity of losing one's job may in fact be so low that no job insecurity is experienced at all.

Thus if employees *do* feel insecure about their jobs, their psychological well-being may be impaired. Conversely, if employees do *not* feel insecure about their jobs, their psychological well-being will not be damaged.

Summary

Drawing on Jacobson's (1985) adaptation of the stress literature, we conceptualized individual reactions to job insecurity. We can expand the framework for studying job insecurity by adding the individual reactions to job insecurity discussed thus far (see Figure 3.2).

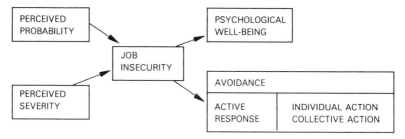

Figure 3.2 *Individual reactions to job insecurity*

Adopting Coping Strategies

Given that avoidance, individual action and collective action are three possible strategies for coping with job insecurity, which conditions make one strategy more attractive than another to a particular individual? More specifically, do characteristics of the person or characteristics of the situation as perceived by the person determine the adoption of coping behaviour? Greenhalgh and Rosenblatt (1984) suggest that personality traits are important moderators between experienced job insecurity and individual responses to that insecurity, and they posit five such traits: locus of control, conservatism, value attributed to work, attribution tendencies and need for security. Jacobson (1985), however, emphasizes the role of cues in the environment. In his treatment of coping with job insecurity, he weighs the perceived benefits of different coping behaviours against the costs of those behaviours.

In our approach we concentrate on two factors we believe are most influential in determining which coping strategy an employee will adopt: the causal attributions employees make to explain their feelings of job insecurity and the perceived costs and benefits of each coping strategy. Greenhalgh and Rosenblatt (1984) mention the first factor, attribution tendencies, as a personality trait that determines individual responses to job insecurity. Like these authors, we hypothesize that those employees who tend to blame themselves for their perceived vulnerability to job loss have stronger responses than those who tend to place the blame on external factors. But, unlike Greenhalgh and Rosenblatt, we define causal attributions not as a personality trait but as the socially constructed meanings employees give to their situation. While we do not deny that personality traits play a role in determining causal attribution, we are more interested in those influences that develop in interaction with the social environment. The second factor, perceived costs and benefits of coping strategies, is discussed by Jacobson (1985). We will expand his discussion by examining the costs and benefits not only of individual but of collective action. Generally, we will argue that causal attributions make strategies more or less pertinent, and that perceived costs and benefits make them more or less attractive.

This introduction raises a number of questions we must answer in order to explain individual differences in responses to job insecurity: (1) What are the different causes to which employees attribute their feelings of job insecurity? (2) What are the determinants of differences in causal explanations? (3) How do causal explanations and perceived costs and benefits of active responses influence an individual's tendency to adopt one coping strategy rather than another?

Causal Attributions

There are many situations in which it is not unusual for people to ask 'Why is this occurring?' and 'Why is it occurring to me?' especially if the situation is one that concerns them deeply. If someone has tried very hard to pass an exam but fails, he may ask himself, 'Why did I fail?' People who survive a disaster often ask themselves, 'Why did I survive?' If one's opponents make a move, one starts wondering why they made that move. What all these observations indicate is that, in order to make sense out of complex reality, people develop their own theories of society and human behaviour. Simple or complex, false or true, these theories are real in their consequences. Each individual responds according to what he or she perceives as the causes of an event. If you believe that the sickness of your child is caused by angry gods taking revenge because you have violated divine laws, you will try to appease the gods rather than go to the doctor for medicine. Or to return to our own subject, if you are convinced that the cause of your job insecurity is poor management, you may try to leave the company rather than enrol in a company retraining programme. For each individual, such causal explanations make certain activities more or less appropriate and thus guide the choices each individual makes.

In social psychology, attribution theory is concerned with the ways in which people try to explain the behaviour of individuals and, more generally, events in their own social environment. Attribution theory has been able to answer some of the key questions about common-sense explanations of life events (Hewstone, 1988). Therefore a summary of the main points of attribution theory may be helpful here. According to Heider (1958), the originator of attribution theory, causal attributions are everyday occurrences that determine much of each person's understanding of and reaction to his or her surroundings. Whenever people ask why something is happening and attribute events to certain causes, they are trying to bring some understandable order into the world in which they live.

It has been argued, however, that people are prepared to let many events occur without even asking why they occur (Eiser, 1987). In responding to such arguments, social psychologists have tried to identify the conditions under which people search for causal explanations. Their studies suggest that attributional activity – that is, activity directed towards finding answers to 'why' questions – is most likely to occur as a response to unexpected events, particularly those involving loss or failure (Weiner, 1985; Wong and Weiner, 1981). Certainly, a change in the future security of one's job is one of those unexpected events which make it likely that people will ask 'why' – 'Why is this occurring, and why is it occurring to me?' The answers employees give to these questions may influence the way they react to job insecurity.

A whole range of questions arises when we examine causal attributions of job insecurity. How do individuals process information and arrive at specific causal explanations? To what extent do workers exhibit forms of defensive attribution – that is, to what extent do they attribute failures (for example, job insecurity) to external causes and successes (for example, job security) to internal causes? To what extent do workers demonstrate ethnocentric attribution, that is, ingroup favouritism (for example, being biased in favour of the workforce) and outgroup denigration (for example, blaming management)? How are intergroup relations – for instance, between the workforce and management – influenced by causal attributions? And to what extent are causal attributions influenced by social ideas shared by large numbers of people within a society (Hewstone, 1988)? In this chapter we will try to answer some of these questions, especially those regarding the processing of information, the occurrence of defensive and ethnocentric attribution, and the influence of shared definitions of the situation. In Chapters 6 and 7, on industrial relations, we will turn our attention to causal attributions and intergroup relations.

Possible Explanations of Job Insecurity
Since we can assume that employees responding to job insecurity will adopt a coping strategy pertinent to the situation as they define it, it is important to know how employees explain the situation in which they find themselves. Do they blame themselves by pointing to their effort, age, health or education? Or do they blame circumstances they themselves cannot be held responsible for, such as poor company management, a declining market for the company's products, or a general economic recession? Are the causes to which they attribute their feelings of job insecurity controllable (for example, one's own effort or a poor management policy) or uncontrollable (for example, one's age or an economic recession)?

As indicated, we may assume that responses to felt job insecurity differ according to the causal attributions one makes. It does not make much sense to look for another job if you hold your age primarily responsible for the uncertain situation you are in. Nor does it make much sense to try to alter a worldwide recession, should that be the factor you hold responsible for the threat to your job. We assume that people select those coping strategies which correspond to the explanations they give for the situation they are in. To put it more specifically, it makes a difference whether employees hold themselves (or some personal characteristic) responsible for their uncertain situation or whether they attribute their condition to forces or circumstances in their environment. Similarly, we can hypothesize that employees will choose differing coping strategies depending on whether they attribute their insecurity to causes that are controllable, either by individual or by collective action, or to causes beyond any control.

From the viewpoint of the individual employee, job insecurity can be caused by internal and/or external factors. Internal factors are characteristics of the individual him- or herself: effort, age, health, education and so on. Some of these factors are controllable, others are beyond an individual's control. External factors can be either intra-organizational or external to the organization. Intra-organizational factors include automation and management policy; factors external to the organization may be recession, governmental policy and a changing demand for the company's products. As a rule, external factors are beyond an individual's control, but some can be influenced collectively, through unions or workers' councils. Since the terms internal and external can be applied from the viewpoint of both the individual and the organization, to avoid any misunderstanding we prefer to describe the three types of causal attribution as individual, organizational and environmental respectively.

Employing the distinction between individual causes, organizational causes and environmental causes on the one hand and controllable and uncontrollable causes on the other, we can classify causal attributions about job insecurity into six different types (see Table 3.1). For each type the explanations we used in our empirical work are given as examples. Education, effort, relationship with colleagues or superiors, work experience and union participation are individual characteristics that can be altered: one can work harder, improve one's level of education and so on. Less subject to control are individual characteristics such as age, health and ethnic background. Causes of job insecurity which are organizational are related to management policy: decisions to close down a department or to reduce the workforce. Though management policy may not always be easy to change, in principle it can be influenced – and sometimes successfully – by collective action through the union or the workers' council. Educational demands and automation, however, are less easy to control. Although organizations can determine the level of education they require from their workforce and the degree of automation they introduce, the extent of their freedom in these matters is usually limited by the requirements of production processes. Policies of institutions external to the company – for example, company headquarters, government or unions – can again in principle be influenced by collective effort. It is more difficult

Table 3.1 *Causal explanations of job insecurity*

	Controllable	Uncontrollable
Individual	education	health
	effort	age
	relationship with colleagues/ superior	ethnic background
	work experience	
	union participation	
Organizational	management policy	educational demands
		automation
Environmental	concern policy	economic situation
	governmental policy	new technologies
	union policy	demand for products

to influence such factors as the economic situation, the development of new technologies and fluctuations in the demand for products.

Sources of Differences in Causal Attribution
If individuals differ in identifying the causes to which they attribute their feelings of job insecurity, it seems reasonable to try to discover the sources of those differences. We can seek them in three different areas: in the objective situation, in the social environment and in personality characteristics.

If the objective situation of individual employees or groups of employees diverges in an observable way, we may expect their causal attributions to differ as well. If, for example, in one company a decline in demand is putting jobs at risk and in another company the introduction of new technologies is the threat, we may probably find that employees in those two companies attribute their feelings of job insecurity to different causes.

The psychological literature has established that individuals differ in their disposition to attribute events to internal or external (Lefcourt, 1976), controllable or uncontrollable causes (Seligman, 1975). In the case of job insecurity, we may expect to find traces of those dispositions in causal attributions people make. For instance, employees who are generally disposed to attribute the events affecting them to external, uncontrollable causes rather than to internal, controllable causes may do the same when confronting job insecurity.

Without underestimating the significance of objective circumstances and personality characteristics, we want to stress that causal attribution is basically a social process, not only because the objects it refers to are social but, more importantly, because attributions are formed, transformed and strengthened in social interaction and because people from the same group or society tend to resort to similar attributions (Hewstone and Jaspars, 1984; Hewstone, 1988).

Because job insecurity is such a complex phenomenon, we must be particularly attentive to the social nature of attribution. As we have stressed earlier, situations that generate job insecurity are in many ways ambiguous; employees may receive cues or hear rumours, sometimes even announcements, that jobs are at risk, but they do not know for certain whether specific jobs (and thus specific individuals) will be eliminated, or when, if at all. Thus uncertainty gives rise to multiple, and divergent, interpretations of the situation. Each person tries to make sense out of the cues he or she is aware of and to give some meaning to the situation as he or she perceives it. On the basis of our knowledge of other social phenomena, we can assume that in situations in which jobs are threatened people do not arrive at their interpretations in a vacuum but in interaction with other people who are in the same situation. In addition, we often see actors such as company management, personnel departments, workers' councils and unions trying to disseminate *their* definitions of the situation among the employees. Moreover, if the events are important enough, newspapers, radio or television will provide their reports and comments. The effectiveness of each actor in convincing employees depends upon its relative significance and its credibility as a source of information for the employees. As a result of this diffusion of information, shared beliefs or 'social representations' may develop among groups of employees (Moscovici, 1981; Hewstone, 1988; Hewstone and Jaspars, 1984; Eiser, 1987). This is not to say that within any group of employees no differences in causal attribution exist. Groups will in all likelihood reach differing degrees of consensus. The extent of mutual interaction within a group and the effectiveness of the various actors in presenting their viewpoints are among the factors we can hypothesize as influencing the degree of consensus within a group of employees.

Causal Attribution and Coping with Job Insecurity

Having introduced the different types of causal explanations of job insecurity, we can now consider our third question: how do differences in causal attribution influence the way people cope with job insecurity? Are avoidance, individual action and collective action equally likely responses in the context of each type of causal explanation? We assume of course they are not, and we will elaborate this hypothesis in this section. A summary of our argument is presented in Table 3.2.

If an employee attributes his or her feelings of job insecurity to *individual/controllable* causes, what coping strategy is he or she most likely to adopt? Attribution to individual/controllable causes is the classic internally controlled situation, which is known to stimulate action-taking (Rotter et al., 1972). For as long as employees see their

Table 3.2 *Causal explanations and related strategies for coping with job insecurity*

	Controllable	Uncontrollable
Individual	individual action	avoidance
Organizational	individual action/ collective action	individual action/ avoidance
Environmental	individual action/ collective action	individual action/ avoidance

job insecurity as related to individual characteristics that they believe are controllable, they can feel there is a way out. They can decide to enrol in retraining programmes, to increase their efforts or to improve their experience. They can look for, and even apply for, other jobs inside or outside the company. Conversely, employees who attribute their job insecurity to *individual/uncontrollable* causes have chosen the most depressing definition of their situation' (Abramson et al., 1978). Not only are they blaming themselves for their insecurity but, equally important, they feel there is nothing they can do. Age, impaired health or ethnic background are factors one cannot change. Therefore, as long as such factors are at the root of an individual's insecurity, he or she may feel powerless. Consequently, a disposition known as 'learned helplessness' easily develops (Seligman, 1975). Under these circumstances avoidance will be the most likely response. We would also predict that this type of causal attribution would have a negative impact on an individual's psychological well-being.

If employees explain job insecurity in terms of *organizational/controllable causes,* what coping strategy will they then adopt? Organizational causes are generally resistant to individual action, but they may be responsive to collective action. For example, employees can try to change management policy by putting pressure on management through the workers' council or the unions. They can even become engaged in industrial action organized by either actor. Of course, such pressure or action is not always successful, but at least success is a possibility. Explanations in terms of *environmental/controllable* causes may similarly generate collective responses. As long as employees believe that actors such as the government, parliament or company headquarters are capable of improving the company's situation, they have definite targets for collective action as well as the hope that their action might succeed. The workforce, sometimes with the support of local management, can pressure these actors to change their policy.

When job insecurity is attributed to *uncontrollable organizational or environmental causes,* it is unlikely that employees will respond with any collective action. What benefits can employees expect from collective action if they believe the situation is uncontrollable? If you feel that a worldwide recession and a declining market are responsible for the problems you are confronting, why engage in collective action? At worst, collective action would cause the company to deteriorate even further. In those situations avoidance is the only response.

This exposition suggests several hypotheses. First, active responses are possible coping strategies for those employees who attribute their job insecurity to controllable causes. Whether those responses will be individual or collective depends on whether employees see the cause in themselves or rather organizational or environmental circumstances. Individual action is a possible coping strategy for those employees who attribute their job insecurity to individual controllable causes. Collective action is associated with controllable but organizational or environmental (social) causes.

Second, avoidance is associated with attribution to uncontrollable causes, be they individual, organizational or environmental.

Third, from the viewpoint of the individual, situations in which job insecurity can be attributed to organizational causes are similar to those attributed to environmental causes, in the sense that they generate the same coping patterns. To be sure, the collective action may differ according to the organizational or environmental origin of the situation: in each case, the action will be directed at different targets, and so the specific strategies the employees will find most effective will also differ. Nevertheless, the two types of situation are similar enough to justify combining organizational and environmental attributions into a single category of social attributions. (In contrast, Chapters 6 and 7 on industrial relations concentrate on organizational versus environmental attributions.)

In sum, we can distinguish four categories of attributions at the individual level. These four categories are presented below, each paired with the coping strategy to which it hypothetically corresponds.

individual/controllable　　– individual action
individual/uncontrollable　– avoidance
social/controllable　　　　– collective action
social/uncontrollable　　　– avoidance

Whether these coping strategies are actually adopted does not depend solely on causal attributions, however. As indicated earlier, causal attribution simply makes a particular strategy more or less pertinent. Actual adoption is also determined by the perceived costs and benefits of each strategy.

The Perceived Costs and Benefits of Action-taking

Once employees have established that action – individual or collective – is a reasonable response to a threat of job loss, how likely is it that any action will indeed follow? What if the costs of taking action are too high? Is an active response equally attractive to each employee for whom such a response seems appropriate? Individual and collective action to preserve one's job can be more or less costly to individual employees. The perceived costs of taking action can range from the time and effort such action will require to increased chances of losing one's job. The perceived benefits range from the satisfaction gained from taking action to increased job security. Moreover, not all employees will arrive at equally favourable estimates of the chance of improving the situation. Because the employees' estimates depend on such factors as their perception of their position in the labour market and their evaluation of their company's situation, they may be very pessimistic about the chances for improving their situation. Among employees who estimate that the costs of taking action are higher than the benefits, avoidance is more likely than action-taking. In other words, it is not only among employees who attribute their job insecurity to uncontrollable causes that we are likely to find avoidance. Employees who estimate the costs of taking action – individually or collectively – to be too high, will be more likely to choose avoidance as a coping pattern even when they attribute their feelings of job insecurity to controllable causes.

Individual Action

Individual responses to job insecurity range from accepting a lower wage, working harder, trying to improve one's relationship with one's superior, gaining further education, to applying for another job. The costs and benefits of such responses involve several elements: What does work mean for a person? How great an effort must be made? and for each possible response, What are the estimated chances of success? Applying the expectancy-value framework (Vroom, 1964; Feather, 1982), we can hypothesize that the more important it is for individual employees to have a job, and the more likely it is in their eyes that a specific individual action will help them keep their current job or provide them with a new job, the more likely it is that they will undertake that action.

The value of work varies with each individual employee, as does the value of specific jobs that are at risk. Understandably, employees who value their present job very highly will tolerate greater costs to make their employment secure than employees who do not really care about their jobs. Similarly, employees for whom it is very important to have work – financially and/or emotionally – will take greater pains to find another job than will employees who are less committed to having work. At the same time, expectations of success also vary among individual employees. Depending on their individual characteristics and on their perception of the labour market inside and outside the company, employees will be more or less optimistic about their chances of keeping their current jobs or of finding other jobs. Note that an employee's perception of his or her position in the labour market has a dual impact in a situation of job insecurity: on the one hand, it makes an individual more or less dependent on his or her present job; on the other hand, it makes the success of an individual action more or less likely.

We can assume that expectations and values have a multiplicative relationship to action-taking; that is, both have to be above zero in order to motivate a person to even try and undertake any action. If, for example, an employee feels that it is not very likely that he will be able to keep his or her current job or find another, the chances are that he or she will give up trying despite the felt importance of having work.

Expectations of success in undertaking individual action are related to individual characteristics such as age, education, ethnic background, work experience and gender which influence one's position in the labour market. Psychologically, such characteristics are reflected in a more or less optimistic view of one's position in the labour market. Thus we may expect older people, people with lower levels of education, foreigners, women and people with little or outdated work experience to be more pessimistic about their chances of finding another job. And the more pessimistic people are about their job prospects, the less likely it is that they will become individually active in response to the job insecurity they experience. On the other hand, the costs of individual action-taking tend to differ for different groups of people, according to the situation they are in. The more committed employees are to their job or their company, the more attached they are to their social environment, the more complicated their financial situation is (it may involve a pension, fringe benefits, and so on), the more costly it will be for them to move to a different job or a different employer. We can hypothesize that employees who are more pessimistic about their chances in the labour market – and/or who are more attached to their present positions – are unlikely to take individual action to reduce job insecurity, except for action that will help them keep their present jobs. Individually, employees can try to keep their present jobs by increasing their effort or by attempting to improve their relationships with their superiors. In other words, one must compete with one's colleagues if one wants to survive. But this competition can cause a rapid deterioration of the social climate in the company (see also Chapters 7 and 8).

Collective Action
Readiness to participate in collective action is far from an automatic response to an adverse situation. Collective reactions to common stressors are rare indeed, and concerted action by an aggrieved collectivity is more exceptional than is often assumed. This is not to say that people are seldom dissatisfied but, rather, that dissatisfaction is not a sufficient stimulus to mobilization. Dissatisfaction must be interpreted in such a way that it can fuel collective protest, and the balance of the perceived costs and benefits of protest participation must be favourable enough to encourage people to take action.

Interpreting Job Insecurity For causal explanations of job insecurity to effect collective action, the workforce must adopt those definitions of the situation which make collective action appropriate (cf. Batstone et al., 1978, and Waddington, 1986, for discussions of vocabularies in support of strike action). This factor brings us back to causal attribution. If job insecurity is attributed to social/controllable causes such as the company management, enterprise management or the national government, collective action is appropriate. Not surprisingly, then, whatever the situation, actors such as unions or workers' councils will attempt to disseminate a definition of the situation in terms of social/controllable causes. It is management that must be improved, government policy that must change, company policy that is wrong, and so on.

Until now, these instances of persuasive communication have received little attention in the literature. As a consequence little is known about the efficacy of unions and workers' councils in disseminating their definition of a situation. We may, however, expect a key issue from the communication literature to apply to the conditions of job insecurity as well, namely, the credibility of a source of information as a determinant of its impact (see Klandermans, 1983, for an illustration in the context of industrial relations). Consequently, we can expect unions or workers' councils to be more influential the more credible they are in the eyes of the workforce.

Perceived Costs and Benefits of Collective Action Let us imagine a situation in which the employees attribute their job insecurity to social/controllable causes, and their union, workers' council or a group of colleagues has tried to persuade them to participate in collective action to pressure the company management to guarantee their jobs in the future. For employees to be persuaded, would it be sufficient for them to see that the goals chosen by those actors coincide with their own concerns about their jobs? Or would it be necessary for them to have some idea of what the chances are that the goals will be realized, and to have some estimate of the risks they run if they participate?

After all, upon reflection the employees might very well find the announced goals unrealistic, and they might justifiably fear that, if they were to participate, they would damage their position in the company – the last risk they would want to take in their uncertain situation. Engaging in collective action, like engaging in individual action, depends not only on the amount of stress people experience but also on expectations of success and perceived costs and benefits of participation. But unlike individual action-taking, collective action involves a more complex relationship between actual behaviour and the likelihood of securing one's job.

To explain this fact we must refer to collective action theory (see, for instance, McAdam et al., 1988). Organizing collective action to secure uncertain jobs generally means that a group of employees tries to prevent dismissals. In other words, the action is usually not aimed at securing the job of a specific person but at achieving guarantees for the collectivity of employees. Further, the collective action may be organized simply to reduce the number of dismissals, the participants knowing that some of them will lose their jobs anyway. In this situation, preventing or reducing dismissals is a collective goal. Whether an individual employee will profit from a decrease in the number of dismissals is only indirectly related to whether he or she participated in the collective actions needed to realize that goal. If the number of dismissals decreases, every employee, independent of his or her participation in any collective action, will benefit. And, to make the situation even more complex, whether a decrease in dismissals will be realized depends only indirectly on the individual's participation. Equally important is the behaviour of others in the same situation. It does not make sense to go on strike or attend a demonstration if no one joins you. A critical mass of participants is needed for collective action to succeed (Oliver and Marwell, 1988). On the other hand, if it appears that the number of participants is great enough to make success likely, an individual may refrain from participation and thus take a free ride.

Under these circumstances it is difficult to convince an individual that his or her participation is crucial. Potential participants themselves realize that the outcome of a collective action depends on what other people do. Consequently, it is sometimes a fairly complicated task to motivate people to participate in collective action. It is not enough to win the employees' hearts for the collective goals; they must also be convinced that there is a reasonable chance that the goals will be realized. For this reason, expectations about the behaviour of others play a crucial role in individual decisions to take part in collective action.

But effectiveness is not the only criterion that affects an individual's decision to participate in collective action. Employees must also

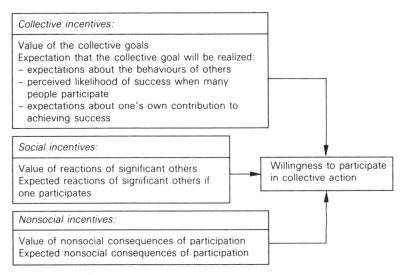

Figure 3.3   *A theory of willingness to participate in collective action*

consider participation in terms of its efficiency, that is, its cost/benefit ratio, both collectively (damage done to the target or to society in general) and individually (risks to be taken and costs to be paid). Organizers must convince employees that the costs of participation stand in a reasonable proportion to the benefits to be gained. The higher the costs, the heavier the burden of legitimation – that is, the more important it will be for the organizers to convince the employees that participation will have some effect. Depending on the kind of activity that is proposed, the costs of participation for an individual can be rather high. These costs may be both social (negative reactions from significant others such as colleagues, family members, supervisors) and material or non-social (loss of income, physical damage, deterioration of one's position in the company). It is reasonable to assume that those employees who fear losing their jobs will be especially concerned about the consequences of their participation. Individual characteristics may also influence specific cost/benefit ratios. Because individual employees differ in their social and financial backgrounds, and their positions in the company, their participation in collective action may be more or less costly.

These factors have been incorporated into a systematic model by Klandermans (1984a and b). This model, developed for the study of individual participation in collective action, can be applied to an analysis of employees' willingness to take part in collective action in response to job insecurity. Figure 3.3 presents the model and its constituent elements. Willingness to participate in collective action is conceived as a function of three types of incentives: collective incentives, social incentives and nonsocial incentives. The strength of each incentive is a multiplicative function of the value of that incentive and the perceived probability that it will be provided. Or to put it more concretely: the value of the collective goal – say, reducing job insecurity – can motivate an employee to participate if and only if he expects the collective action to succeed in realizing that goal. Expectations of success for collective action can – theoretically – be broken down into three components: the individual's expectations about the behaviour of others, the perceived likelihood of success when many people participate, and expectations about one's own contribution to the achievement of success. Each component must be above zero for the value of the collective goal to motivate participation. Similarly, the expected reactions of significant others and expected nonsocial consequences can influence a person only if he or she cares about those reactions or consequences.

Summary

We have conceptualized job insecurity as a function of the perceived probability and the perceived severity of losing one's job. In answering our first question, 'What makes employees feel insecure about their jobs?' we hypothesized that characteristics of the organization and the industrial-relations climate in the organization, individual characteristics of the person, his or her position in the organization, and personality characteristics affect the perceived likelihood of losing his or her job. The perceived severity of losing one's job is, we assumed, influenced by the value of particular job features and by one's dependence on one's present job for the acquisition of important job features.

In answering our second question, 'How do employees react to feelings of job insecurity?' we distinguished between the impact of job insecurity on psychological well-being and coping with job insecurity. In general, it was expected that feelings of job insecurity would lower the level of psychological well-being. Coping with job insecurity was described in terms of two basic strategies: avoidance – psychological withdrawal – and active response – attempts to restore security. Attempts to restore security were subdivided into individual active responses and collective active responses.

Our third question, 'What explains individual differences in responses to job insecurity?' we answered in terms of the causal attributions employees make in their attempts to explain the insecurity they experience and also in terms of the perceived costs and benefits of each coping strategy.

4

Predicting Employees' Perceptions of Job Insecurity

Tinka van Vuuren, Bert Klandermans, Dan Jacobson and Jean Hartley

Job insecurity, then, is a subjective phenomenon. Employees reading the cues in their environment for some reason begin to fear for the continuity of their jobs. The key questions of course are: what cues make people worry about the future of their jobs and why? As we will see in this chapter, there are no simple answers to these questions. Although the cues that trigger feelings of job insecurity may indeed be present in a given situation, some employees will observe them, but others in the same situation will not. Moreover, cues that worry one employee may provide another with reason for optimism. For example: in the British study, the start of the production of a new vehicle was for some employees a reason to fear new dismissals, whereas others found it reassuring. In other words, a situation that makes some employees feel insecure about the future of their jobs may make others feel entirely secure – sometimes even for the same reason. Neither of these diverging perceptions of a situation is in itself right or wrong. Even among our own small number of cases, we found that history sometimes justifies those who were concerned, sometimes those who were not. Thus, in the Israeli case, the optimists got it right: ultimately, nothing happened. But in the British case it was the pessimists who were proven right: some time after the last interviews were conducted the company went into receivership.

This is not to say that feelings of job insecurity appear randomly or that they are impossible to predict. The theoretical framework for the prediction of job insecurity which we developed in Chapter 3 indicates that, on the contrary, job insecurity is indeed a predictable phenomenon, and it is to an empirical test of this framework that we now turn. We will begin by establishing the extent to which individuals employed in the same organization differ in their feelings of job insecurity, and then consider possible explanations for divergent appraisals of a similar environment. Given the cross-sectional data, we cannot test our model as a causative model. Although the cross-sectional relationships revealed may suggest some conjectures about causation, specific research into causation is needed to put causality to a test. On the other hand, in many places we can reasonably assume that causality is bi-directional.

We noted earlier that the three studies that form the basis of this book were not specifically designed for rigorous comparison. Some concepts from the framework outlined in the previous chapter were included in each study, sometimes operationalized identically, sometimes slightly differently.

The Experience of Job Insecurity

Let us suppose that the government announces it will reduce the number of civil servants, or that the company management warns that the organization must become more efficient, or that it becomes known that the company's order book is dangerously empty, would the affected employees fear for the future of their jobs, or would they for some reason feel that the situation will not deteriorate too much? It seems only logical that they would be concerned, but how many do in fact feel insecure under such circumstances?

The studies conducted in Israel, Great Britain and the Netherlands provide answers to these questions. As we mentioned briefly in Chapter 1, the respondents of the three studies were employed, in Israel, by the Ministry of Labour and Social Welfare, after the government had announced cutbacks; and, in Great Britain and the Netherlands, by manufacturing companies in a state of decline. Figure 4.1 presents the percentages of employees' feelings of job insecurity from these organizations. Employee responses are categorized on a scale ranging from 'one' – very secure – to 'five' – very insecure. It is misleading to compare the data in Figure 4.1 between studies. Measures are not exactly identical, and circumstances, history and context differ too much to permit any relative conclusions. Therefore, we will not pursue possible explanations for the differences between countries.

More important for our present discussion are the percentages of employees experiencing more or less job insecurity within each study. In the Israeli, Dutch and British studies alike, we found substantial proportions of both insecure and secure employees. Apparently, people in the same organizational circumstances can differ considerably in their evaluation of a given situation. Consider, for example, the following quotations from two employees in the British study: one responded, 'The joke around here is – will the factory be open when I reach the main gate in the morning? That's what a lot of people think'; the other claimed, 'The company, as you know, is on the up. I haven't been concerned about the security of my job for a couple of years now.' It is hard to believe that these two people were talking about the same company.

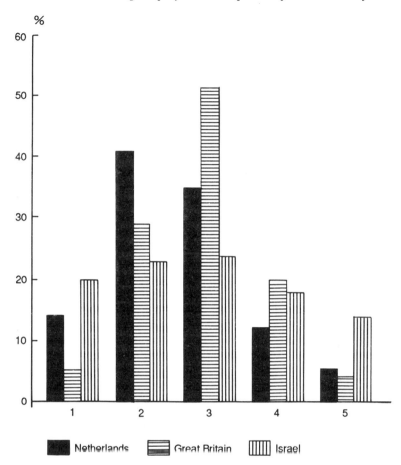

Figure 4.1 *The experience of job insecurity*

Results from the Dutch corroborate the argument that people can differ considerably in their evaluation of the situation. After asking the Dutch respondents whether they were concerned about the future of their job, we tried to identify the reasons underlying their appraisals of the situation (see Table 4.1). We distinguished between individual reasons such as education, health, effort and experience, and social reasons such as management policy, economic situation and new technologies. Interestingly, in the case of social reasons, the same cues were cited by both secure and insecure employees. For example, while management policy is a source of concern for the latter, for the former it is a source of assurance. Similarly, automation or the implementation of new technology makes some employees fearful, others hopeful. Clearly, even when referring to the same cues, different people can arrive at opposite conclusions.

Table 4.1 *Reasons for concern or unconcern about job insecurity in the Dutch study*

	Reason to be concerned % (n = 150)	Reason to be unconcerned % (n = 158)
Individual		
Education	14.0	87.9
Work experience	4.0	94.9
Own effort	3.3	93.0
Relationship with colleagues/superior	13.4	95.6
Activities for union	7.3	15.8
Employees' efforts	32.7	83.5
Health	10.0	89.3
Age	52.0	54.2
Ethnic background	12.1	36.1
Extra-individual		
Management's decision-making	57.0	65.6
Management's efficacy in opening new markets	69.3	45.8
Enterprise policy	64.0	58.2
Governmental policy	67.3	44.5
Union policy	36.5	62.5
Demands for products	80.0	72.1
Economic situation	77.3	54.2
New technology	58.0	68.8
Automation	42.3	67.0
Educational demands	60.8	73.9

But perhaps even more interesting are the results concerning individual reasons. Employees who are not concerned about the future of their jobs overwhelmingly refer to individual characteristics as the reasons for their lack of concern. Conversely, employees who feel insecure rarely mention individual characteristics as a reason for their concern. The British study produced results comparable to those yielded by the Dutch survey. In this study, the workforce was asked to explain previous redundancies. Explanations focusing on one's own party were not perceived as very significant: only 20 percent, for example, endorsed the view that the workers' lack of effort contributed to previous job losses, and an even smaller percentage cited other causes based on one's own party. Poor decisions by plant management were seen as the most important cause of job losses in the past. This pattern of attributions is typical of intergroup processes between parties involved in conflict or in a situation of mistrust: the other party is the object of negative attributions, while one's own party is perceived in a positive light.

To summarize: even when employees are in identical organizational circumstances and refer to the same cues, they may arrive at different conclusions about their future prospects. If the same circumstances and cues produce different responses, we must explore other predictors that might explain feelings of job insecurity.

Measuring Predictors of Job Insecurity

Since job insecurity is a perceptual phenomenon, the predictors of job insecurity must specify aspects of the reality perceived by the individual employee, such as trust or distrust in management or a supposed indispensability. Some predictors of job insecurity are factors that the individual believes will make job loss more likely; other predictors are factors that would make the loss of one's job more difficult to bear. We can predict that the more likely *and* the more serious it would be for employees to lose their jobs, the greater their concern about the future of their jobs. In Chapter 3 we discussed four factors that predict the likelihood of losing one's job: the organizational and industrial-relations climate, individual characteristics and characteristics of one's position in the company, and personality characteristics. We also examined several predictors of the severity of losing one's job: the range of job features that could be in jeopardy, the valence of each of those features, and the employee's dependence on the present job for each feature.

The perceived organizational and industrial-relations climate encompasses beliefs about management, the unions and the workers' council. The basic question is: are management, the workers' council and the unions able and willing to protect employees from losing their jobs? To determine the role of management, we operationalized this question in terms of trust in management. In the British and Dutch studies, trust in management was measured by a six-item Likert-type scale derived from the Interpersonal Trust-at-Work Scale of Cook and Wall (1980). In both studies the items were translated into terms, not of interpersonal trust, but of trust between management on the one side and the workforce and unions (workers' council) on the other. The scales consisted of such statements as 'Management can be trusted to make sensible decisions for the plant's future' and 'Our firm has a poor future unless it can attract better management.' Because of different industrial-relations institutions in the two countries, the scales in the two studies were not completely identical. Nevertheless, the scales can serve very well in helping us to compare the two studies' data on the significance of trust in management for job insecurity. In the British study, levels of trust in management were low for the group as a whole. To illustrate: 65 percent of the respondents felt that management would be quite prepared to gain the advantage by deceiving the unions; 53 percent disagreed with the statement that 'Management can be trusted to make sensible decisions for the plant's future'; and 66 percent agreed that the company would have a poor future unless it could attract better managers. The Dutch employees were much less negative about management. Twenty percent of the Dutch respondents thought that management would be prepared to deceive the workers' council; 25 percent trusted management to make sensible decisions for the company's future; and 36 percent felt that the company would have a poor future unless it could attract better management.

Opinions about the workers' council and the unions were operationalized in terms of the perceived strength of the workers' council and the union. In the Dutch study, this perceived strength was measured by two Likert-type scales including items such as 'The workers' council (union) has little influence in the company.' The Dutch employees deemed the unions to be slightly more influential than the workers' council: for instance, 40 percent disagreed with the opinion that the union has little influence in the company, compared to 34 percent in the case of the workers' council. In the British study, we used a four-item Likert-type scale of perceived strength of the union based on a scale used in a Dutch study of union participation (Klandermans, 1983). This scale included statements such as 'There is a lot that unions can do to change what goes on in the workplace' and 'It doesn't make sense to fight for your interests since workers have very little power.' In the case of Israeli employees, any investigation of scales relating to the perceived strength of a union is complicated by factors relevant to the specific position of the General Federation of Labor (Histadrut) in Israel. Therefore, in our Israeli study, scales related to industrial relations were restricted to the perceived strength of the workers' council.

A variety of *personal characteristics and characteristics of the employee's position in the company* can influence an individual's perception of the likelihood of losing his or her job. In both the Israeli and the Dutch studies these characteristics were regarded as perceived safeguards, as attributes that may protect employees from losing their jobs. Both studies used almost identical lists (see Table 4.2). Individual employees were asked whether they felt that one or more of these attributes protected them against job loss. The two studies reveal similar patterns: in both cases personal characteristics – for example, personal output, capabilities and work experience – were mentioned as safeguards more often than were characteristics of one's position in the company – seniority, for example, or importance of one's job, or indispensability. This response confirms our earlier finding that employees who were not concerned about the future of their jobs more often referred to individual characteristics in explaining their lack of concern, than to positional characteristics.

Table 4.2 *Perceived safeguards against job insecurity*

	Israel % (*n* = 233)	Netherlands % (*n* = 310)
Seniority	33.9	19.7
Importance of one's job	60.0	–
Importance of one's department	64.4	–
Indispensability for the company	–	17.1
Family situation	11.6	8.1
Health	8.1	28.7
Connections 'upstairs'	14.6	20.0
Direct supervisor	45.1	28.5
Personal output	71.2	45.8
Ethnic background	2.1	11.3
Professional capabilities	70.4	41.4
Popularity among colleagues	40.0	24.2
Work experience	–	44.8
Special skills	59.2	–
Activities for the union	–	2.0
Workers' council	17.1	12.6
Trade unions	7.7	16.1

In the Israeli study the personal and positional characteristics yielded two factors explaining 52 percent (eigenvalue 5.80) and 23 percent (eigenvalue 1.48) of the variance respectively. The first factor – containing the attributes seniority, connections, supervisor, workers' council and importance of one's job/department – was labelled 'positional safeguards'; the second – containing the attributes personal output, professional capabilities and special skills, and popularity among colleagues – was labelled 'personal safeguards'. In the Dutch study, factor-analysis of personal and positional safeguards produced a single factor explaining 80 percent of the variance (eigenvalue 5.11). This factor was labelled 'perceived safeguards'. We constructed variables indicating to what extent the individual considered the perceived safeguards applicable in his situation. Thirty percent of the Dutch respondents felt that none of these safeguards was applicable in their situation.

The employee's tendency to be more or less concerned about the future of his job has been related in our studies to three *personality characteristics*: locus of control, self-esteem and pessimism. Locus of control (mean score 3.31, SD 0.76) and self-esteem (mean score 1.74, SD 0.90) were administered in the Israeli study; locus of control (mean score 2.90, SD 0.53) and pessimism (mean score 1.44, SD 0.27) in the Dutch study. Abbreviated versions of existing personality measures were administered. External locus of control, low self-esteem and a generally pessimistic outlook were thought to increase feelings of job insecurity.

We included a list of *job features* in the Dutch study to find out which job features were perceived to be at risk. These features are, with a few additions, adopted from Greenhalgh and Rosenblatt (1984): career progress, status, work content, income, autonomy, responsibility, job-specific work experience, social relations and workload. Income and work content were perceived to be at risk by two-fifths of the respondents; career progress by one-fifth. Only small percentages of the employees mentioned the other features. Fear of losing these job features was significantly correlated to feelings of job insecurity, although some features, such as concerns about career, status, work content and income, were more strongly correlated than others, for example, workload and social relations. From this list we constructed a variable indicating the range of job features thought to be at risk.

Financial vulnerability exacerbates the severity of losing one's job. If an individual's current job provides a high proportion of the family income, or if an individual faces high fixed obligations, the prospect of losing his or her job is especially threatening. In the Dutch study, 49 percent of the respondents felt that their financial position made them vulnerable. In the Israeli study, the respondents' subjective sense of their financial positions was assessed by asking whether they were economically better or worse off than people in comparable positions: 18 percent of the Israeli respondents felt they were worse off. In the British study, the number of dependants and the number of wage-earners in a household were assessed.

Following Greenhalgh and Rosenblatt (1984), we assumed *dependence on the present job* to consist of the employee's perceived position in the labour market. Skills and professions that are in low demand, as well as personal characteristics that put people in an unfavourable position in the labour market (for example, age, health, gender and ethnic background), make employees more dependent on their present jobs and intensify the consequences of job loss. In the Dutch study, this dependence variable was measured subjectively by asking respondents whether their position in the labour market was unfavourable. Half of the respondents felt that their positions were unfavourable.

In addition to these perceptual data, we included in the analyses such variables as seniority, age, income, education, marital status and union involvement. Because there were very few female respondents included in the studies, we did not treat gender as a separate variable.

Table 4.3 *Antecedents of job insecurity: Pearson correlations
with feelings of job insecurity*

	Israel ($n = 233$)	Netherlands ($n = 308$)	UK ($n = 137$)
Factors influencing perceived probability			
(a) organizational/industrial-relations climate			
– trust in management		$-.39^3$	$-.33^3$
– perceived strength of workers' council	$-.18^1$	$-.18^2$	
– perceived strength of unions		$-.16^2$	$-.23^2$
(b) personal/positional characteristics			
– perceived safeguards		$-.48^3$	
positional safeguards	$-.40^3$		
personal safeguards	$-.23^2$		
(c) personality characteristics			
– locus of control (more external)	$.19^2$	$.29^3$	
– self-esteem	$-.16$		
– pessimism		$.32^3$	
Factors influencing perceived severity			
(a) – job features (more features involved)		$.42^3$	
– financial position			
perceived financial vulnerability		$.12^1$	
number of dependants			$.16^1$
number of wage-earners			$-.17^1$
relative economic status	$.18^2$		
(b) dependence on present job			
– perceived position in labour market (stronger)		$-.22^3$	
Demographics			
– education		$-.14^1$	
– age	$.03$	$.04$	$.18^1$
– seniority			$.12^1$

$^1 \ p < .05.$
$^2 \ p < .01.$
$^3 \ p < .001.$

Predicting Job Insecurity

To what extent are the predictors outlined in our model and described
in the previous section capable of explaining the variance of job
insecurity? Table 4.3 shows the Pearson correlations between job
insecurity and the predictors together with the demographic factors.
Correlations not included in the table could not be computed, because
all predictors were not always included in a specific study. Since the
British study focused primarily on the consequences of job insecurity
on plant-level industrial relations, fewer predictors were included in

this study. Correlations with those predictors that were included in the British study are of course displayed.

The correlations in Table 4.3 confirm the hypotheses that were implied by our theoretical model. We found the following relations:

1 The greater the employees' trust in management, and the stronger the workers' council and unions are perceived to be, the more secure employees feel about the future of their jobs.
2 The more safeguards employees believe they possess, the more secure they feel about their jobs.
3 The more employees believe that their situation is externally controlled the less self-esteem they have; and the more pessimistic their outlook, the greater their insecurity about their jobs.
4 The more job features employees believe to be endangered, the greater the degree of job insecurity they report.
5 Employees who feel they are in a weak financial position feel more insecure about their jobs than do employees who feel their financial state is fairly strong.
6 Employees who believe they have a strong position in the labour market feel more secure about their jobs than do employees who feel more vulnerable.
7 More highly educated employees in the Dutch study, and younger employees in the British study, feel more secure about their jobs.

These findings require a few comments. First, it is important to notice that most findings could be produced in more than one country. This fact underscores the potential for generalizing our findings. Second, except for small but significant correlations with education and age, job insecurity was virtually unrelated to demographic variables. Age, seniority, education, ethnic background and so on were included in the list of safeguards, however, and as such formed part of the strongest correlate of job insecurity. This finding is especially interesting, for it suggests that objective factors such as age, health, education and seniority are filtered through a process of cognitive appraisal. If and only if employees *feel* that their age, education or ethnic background makes them more vulnerable do these factors increase job insecurity. Moreover, in general, there is no relationship between job insecurity and location (department or function) in the organization. The only exception to this generalization appeared in the Dutch study: employees who worked in a department that had experienced job losses in the recent past felt more threatened by present circumstances than did employees from other departments in the same company. The absence of a definite relationship between insecurity and position in the company in situations of job insecurity in general becomes understandable when we consider the influences that shape an employee's perceptions. First, all these employees were in the same organizational

Table 4.4 *Multiple regression of perceived probability of losing one's job on antecedents: Israeli study*

	Standardized beta
(a) Intra-organizational safeguards	$-.47^3$
(b) Intra-individual safeguards	$-.14^2$
(c) Locus of control	$.12^1$
(d) Perceived strength of workers' council	$-.12^1$
(e) Self-esteem	$-.11^1$
r^2	.40

^{1} $p < .05$.
^{2} $p < .01$.
^{3} $p < .001$.

situation. In each case, it was a department, a plant or the workforce as a whole that was affected. In such a situation it did not matter what job an individual held or in what department he or she worked. Nevertheless, some employees persisted in thinking that their particular job or department was too important to be eradicated. But we found that even among people who work in the same department or who have the same function, such confidence was not universal.

Finally, we can return to our central question: how well do these predictors account for the variance in job insecurity? We applied multiple regression analyses to assess the relative importance of several predictors together in explaining job insecurity. Because fewer predictor variables were included in the British study, only the Israeli and the Dutch studies were subjected to these analyses. Table 4.4 focuses on the perceived probability of losing one's job and on the factors influencing this dimension of job insecurity which were included in the Israeli study. These variables could explain 40 percent of the variance in the perceived probability of losing one's job, as we can conclude from the R^2. Each of the variables included in Table 4.4 contributes significantly to the perceived probability of losing one's job, as the level of significance of the single betas indicates. Among these variables, perceived safeguards were the strongest predictors, as indicated by the size of the beta ($-.47$ in the case of intra-organizational safeguards and $-.14$ in the case of intra-individual safeguards; together these two variables account for 36 percent of the variance). In the Israeli study, the perceived strength of the workers' council and the two personality characteristics in the Israeli study (locus of control and self-esteem) are about equally important.

Table 4.5 presents the results of feelings of job insecurity regressed on the predictors included in the Dutch study. The results for the factors influencing the perceived probability confirm those from the Israeli study, with one intriguing exception, the positive sign of the beta for the perceived strength of the workers' council.

Table 4.5   *Multiple regression of job insecurity on antecedents:*
Dutch study

	Standardized beta
Factors influencing perceived probability	
(a) perceived safeguards	$-.38^4$
(b) trust in management	$-.35^4$
(c) pessimism	$.14^3$
(d) locus of control	$.12^2$
(e) perceived strength of workers' council	$.15^2$
Factors influencing perceived severity	
(f) job features at risk	$.13^2$
(g) perceived position in labour market	$-.09^1$
r^2	$.45$

^{1} $p = .07$.
^{2} $p < .05$.
^{3} $p < .01$.
^{4} $p < .001$.

Together, the variables included in Table 4.5 explained 45 percent of the variance in job insecurity among the Dutch employees. Twenty-five percent of the variance explained can be attributed exclusively to the predictors that influence the perceived probability, 2 percent exclusively to those influencing the perceived severity of losing one's job, and 18 percent to the joint sets of predictors. These percentages indicate that the predictors influencing the probability of losing one's job were far more significant than the predictors influencing perceived severity. This result reinforces the finding reported by Ashford et al. (1987) discussed in Chapter 2: powerlessness was the most important correlate of job insecurity, that is, the inability to have any influence on the likelihood of losing one's job, in their study.

In the Dutch study as in the Israeli study, perceived safeguards turned out to be the most important determinant of feelings of job insecurity. Also in the Israeli study, personality factors were important: the impact of locus of control is similar in the two studies; in addition, pessimism appears to have a significant influence.

The other major factor in the Dutch study is trust in management. In the Dutch data, trust in management is strongly related to the perceived strength of both the workers' council and the union ($r = .67$ and .58 respectively). It appears that the three factors represent the employees' appreciation of the industrial-relations climate within the company. That is, in expressing their trust in management, employees also expressed their belief in the efficacy of the employees' council and unions. What influences employees' feelings of job insecurity, then, is their conviction – or doubt – that these actors in mutual collaboration are capable of averting threats to the continuation of their jobs. Of particular interest is the finding that the perceived strength of the workers' council is positively related to job insecurity after we have controlled for trust in management (the same relation holds for the perceived strength of the unions, but this variable is no longer important after the perceived strength of the workers' council has been entered). Apparently, the perceived strength of these two parties helps to reduce feelings of job insecurity only in conjunction with trust in management.

The factors influencing the severity of losing one's job contribute significantly to the explanation of job insecurity. Of these factors the number of job features perceived to be at risk is the most important. The perceived weakness of one's financial position did not add significantly to the variance explained. Perceived position in the labour market, however, almost reaches significance.

Summary

In both the Israeli and Dutch studies, the perceived safeguards against job loss turned out to be the most important determinant of feelings of job insecurity. This finding demonstrates that if people possess some attribute they think to be protective, they will use it to reason away the threat and so reduce their uncertainty. One implication of this finding is that unions, workers' councils and management must each draw a realistic picture of the company's situation. Such realism may on the one hand reassure employees who erroneously fear that their jobs are in danger; on the other hand, it may also open the eyes of those who mistakenly believe that particular safeguards will keep them out of danger.

The second most important factor was the appreciation of the industrial-relations climate in the company. People feel much more secure about their jobs when they can depend on strong unions and on a strong workers' council that has a responsive management to deal with, a management perceived as capable of averting threats to the employees' positions. An important precondition for avoiding unnecessary job insecurity, then, is to create and preserve a climate of trust. Measures that raise questions about management's good intentions and/or capabilities of management can rapidly undermine the industrial-relations climate, creating an accelerating downward spiral of mistrust and deepening insecurity.

The significance of personality characteristics in relation to feelings of job insecurity reminds us that interpreting environmental cues involves a strong personal factor. People who tend to be pessimistic in general will interpret the situation in their work environment pessimistically as well. This is not to say that they create problems that do not exist; rather they are especially sensitive to the warning signals in their environment. Here an observation of Taylor and Brown (1988) is particularly relevant: depressive people will evaluate a situation more realistically than will non-depressive people.

5

Employees' Reactions to Job Insecurity

Tinka van Vuuren, Bert Klandermans, Dan Jacobson and Jean Hartley

When employees fear for their jobs, how do they react? Does insecurity about the future of their jobs make them more depressed and apathetic or just the opposite – motivated to seek another job or to become engaged in industrial action? And, given these different options, why do they choose one rather than another?

In analysing the individual's responses to job insecurity we have distinguished psychological well-being, avoidance, individual action and collective action. These responses are thought to be generated by two processes: that of causal attribution and that in which the employee weighs the perceived benefits of action-taking against the perceived costs. In the first part of this chapter, we will investigate whether the psychological well-being of employees is affected by job insecurity and whether they choose to withdraw psychologically from their situation or to respond actively. As we will see, employees differ considerably in their responses to job insecurity.

The second part of the chapter tries to explain why individuals differ so greatly in their reactions to job insecurity. Why are some people apparently untouched by feelings of job insecurity while others become depressed? Why do some people resort to psychological withdrawal while others take individual or collective action to restore job security? We presume that these differences stem from divergent causal attributions and cost/benefit evaluations of action-taking.

Responses to Job Insecurity

In this section we will compare insecure employees with secure employees in relation to several characteristics: well-being, psychological withdrawal and behaviours such as retraining, job-seeking and union participation.

Mental Health and Work Satisfaction

Psychological well-being is an umbrella that stands for a range of emotional and cognitive states. An individual's mental health, happiness, satisfaction with life or with work are all considered aspects

Table 5.1   *Job insecurity and mental health*

	Israel			Netherlands		
	High (n = 160)	Low (n = 73)		High (n = 135)	Low (n = 173)	
Depressive affect	12.46	8.32	F = 48.77[2]	8.39	5.15	F = 40.14[2]
Psychosomatic complaints	2.56	2.15	F = 8.45[1]	1.95	1.64	F = 7.93[1]

Analysis of variance
[1] $p < .01$.
[2] $p < .001$.

of his or her psychological well-being. Does job insecurity in fact have the negative impact on psychological well-being which we hypothesized? Does it make people feel less happy, less satisfied? Does it impair their mental health? In our surveys, we investigated several indicators of psychological well-being, focusing in particular on two aspects – mental health and work satisfaction. The two following sections present findings on each topic in turn.

Mental Health   The Israeli and the Dutch studies both included several measures of mental health. Specifically, both studies used the Depression Affective Adjective Check List (Lubin, 1967), the Israeli study in the Hebrew translation by Lomranz et al. (1981), the Dutch study in the translation by Van Rooijen (1984). Lubin's depression measure consists of a list of affective adjectives, such as 'unhappy', 'active', 'lost', 'concerned', 'fine', 'excellent', 'hopeless' and 'lonely'. Respondents must indicate whether, or which, specific adjectives apply to their feelings. In the Dutch study participants were given in addition a scale derived from the Dutch Affect Check List (Hoekstra, 1986). This scale consists of a list of affective reactions. Respondents were asked to indicate for each reaction whether it matched one of their own responses when they thought about the future of their jobs. The Israeli study included a series of questions on psychosomatic complaints, which were derived from Warr (1987) such as: 'Have you had difficulty falling asleep in the last few weeks?' 'Have you been quite nervous?' 'Have you had headaches?' The Dutch survey posed a single question about such complaints, asking whether respondents have been troubled by sleeplessness, headaches, tension or fatigue during the last few weeks. Table 5.1 presents the results of analyses of variance from both the Israeli and Dutch studies with depressive affect and psychosomatic complaints as the dependent variables and job insecurity as the independent variable (the high and low levels of job insecurity were determined by the mean). Table 5.2 reports the results from

Table 5.2 *Affective reactions to job insecurity: Dutch study*

| | Perceived job insecurity | | |
	High ($n = 135$)	Low ($n = 173$)	
Nervousness	3.17	2.36	F = 63.33[2]
Anger	2.98	2.31	F = 46.82[2]
Guilt	2.02	1.83	F = 6.69[1]
Pleasure	2.55	3.26	F = 38.52[2]
Sadness	2.97	2.18	F = 60.83[2]
Assertion	2.85	3.41	F = 38.74[2]
Fear	2.84	2.08	F = 59.38[2]

Analysis of variance.
[1] $p < .01$.
[2] $p < .0001$.

analyses of variance from the Dutch study, again with job insecurity as the independent variable and the various affective reactions as dependent variables. Table 5.1 demonstrates that in both studies job insecurity is related to poorer mental health. Employees who felt very insecure about their jobs had more psychosomatic complaints and were more depressed than employees who felt secure about their jobs. We cannot compare data between countries because the measures we used and the circumstances we surveyed were too different, but both studies reveal the negative impact of job insecurity on mental health. Table 5.2 confirms the picture outlined in the preceding table. Employees from the Dutch study who felt insecure about their jobs reported more nervousness, guilt, sadness, fear and anger, and less pleasure and self-confidence than employees who did not feel insecure.

Work Satisfaction Both the British and the Dutch studies investigated work satisfactions. In both studies we examined both job satisfaction and organizational commitment. The scales derived from items from existing measures (Warr et al., 1979; Cook and Wall, 1980 respectively). Using the satisfaction scale, we assessed, in the Dutch study, the extent to which employees were satisfied with such aspects of their job as pay, the amount of work, supervision, career perspectives, work content, autonomy, responsibility and so on. The British study examined pay, amount of work and supervision. The organizational-commitment scale was composed of six statements such as 'I am quite proud to be able to tell people I work in this organization' and 'The offer of a bit more money from another employer would seriously make me think of changing my job.' Table 5.3 presents the Pearson correlations of both measures with job insecurity in the British and Dutch studies. Both studies yield significant correlations between job insecurity and

Table 5.3   *Pearson correlation of job insecurity with job satisfaction and organizational commitment*

	UK	Netherlands
Job satisfaction	$-.43$	$-.32$
Organizational commitment	$-.31$	$-.37$

All correlations are significant at $p < .01$.

work satisfaction. It is clear that these results lead to similar conclusions: feelings of job insecurity are accompanied by lower job satisfaction and a weaker commitment to the organization.

Avoidance, Individual Action and Collective Action
While some people react to job insecurity by withdrawing from their work psychologically, other people choose to attempt to restore their job security either by individual or by collective action. In our theoretical discussion we distinguished avoidance, individual action and collective action as three strategies an employee might use to cope with job insecurity. But we must still demonstrate that these three coping strategies actually occur in response to feelings of job insecurity. Do employees who feel insecure about the future of their job indeed withdraw psychologically from work, or do they undertake individual and/or collective action to restore job security?

Table 5.4   *Coping strategies*

	Perceived job insecurity					
	Israel			Netherlands		
	High ($n = 160$)	Low ($n = 73$)		High ($n = 135$)	Low ($n = 173$)	
Avoidance	–	–		2.1	1.9	$F = 8.62^2$
Individual action	3.6	3.5	$F = 1.48$	1.5	1.2	$F = 16.73^3$
Collective action[1]	2.6	2.1	$F = 9.84^2$	3.3	2.8	$F = 10.92^3$

Different scales are used in Israel and the Netherlands.
Analysis of variance.
[1] In Israel based on a scale encompassing ten different forms of collective action in varying degrees of militancy; in the Netherlands on a measure encompassing willingness to participate in moderate and militant collective action.
[2] $p < .01$.
[3] $p < .001$.

Table 5.4 presents results from the Israeli and the Dutch studies. In the Israeli study no separate measure of avoidance was included. Nevertheless, the general picture is clear: Dutch employees who felt

Table 5.5 *Responses in job insecurity: Dutch study*

	(%) Perceived job insecurity	
	High (n = 135)	Low (n = 173)
Avoidance		
Feels little motivation to go to work	34.0	17.3
No interest in one's work	29.1	8.7
Doesn't bother about one's work	61.4	55.5
No dedication to one's work	12.6	5.8
No interest in the situation of the company	34.8	20.2
Doesn't talk about the situation of the company	6.7	16.2
Individual action		
Seriously thought about seeking another job	26.7	11.6
Paid much more attention to information about possible other jobs	22.2	9.2
In fact applied for another job	23.1	12.9
Collective action		
Willing to participate in moderate action	78.5	56.3
Willing to participate in militant action	43.7	22.7

insecure about their jobs displayed more signs of avoidance; Israeli and Dutch employees who felt insecure about their jobs were more often engaged in individual and/or collective action to restore job security than employees who were not insecure. Referring to both the Dutch and the Israeli studies for examples and evidence, we will now present more detailed information on each of the three coping strategies.

Avoidance. As the percentages from the Dutch study indicate (see Table 5.5), some employees who experience high levels of job insecurity withdraw psychologically from their work. These employees feel less motivated to go to work, they are less interested in their work and they are less dedicated to their work than are employees who feel less insecure. Moreover, they are less interested in the situation of their company as a whole. Psychological withdrawal does not mean, however, that the situation of the company is no longer a topic of conversation. On the contrary, employees who feel insecure about their jobs discuss the company's situation more often than do employees who are not insecure. To this extent, avoidance as a coping strategy differs from denial of the problem, a condition in which we would expect that employees do not even talk about the problem.

Individual action. Employees who experience a high degree of job insecurity are more willing to undertake individual action more often than employees who do not feel threatened. The data from the Dutch study in Table 5.5 demonstrate that they will consider seeking another job more often than their more secure colleagues do; they pay much more attention to information about other possible jobs and they apply for other jobs more often.

Using the Israeli data, we can elaborate in more detail the information-seeking behaviour of insecure employees. In the Israeli context, networking – that is, consulting friends and contacting potential employers about other possible jobs – is in particular a strategy more often adopted by people who feel highly insecure about their jobs than people who do not. Another possible response to job insecurity is to increase the effort one puts into one's work. This reaction has been described in the literature on survivors of collective lay-offs (Brockner et al., 1985). The Dutch study did not yield any evidence of such a reaction, but the Israeli study did. Seventy-three percent of the employees experiencing high levels of job insecurity admitted that they were putting a much greater effort into their work than usual, while 56 percent of those who did not experience job insecurity put more effort into their work than usual. It is interesting to note that employees who felt that job losses were likely to reflect on personal ability and performance were those most likely to put greater effort into their work. This finding confirms our hypothesis that reactions to job insecurity depend on the definition of the situation adopted by the individual.

Collective action. Feelings of job insecurity did not alter employees' involvement in the union in any of the three countries: in Israel, Great Britain and the Netherlands union membership, activity levels and levels of union commitment proved to be unrelated to concern about the future of one's job (see also Chapter 7).

An employee's willingness to participate in industrial action, however, *is* influenced by feelings of job insecurity, as the data from each of the three countries illustrate. Attitudes towards militant actions such as strikes and occupations provide the most convincing evidence of this effect.

Through two questions – one on strikes and the other on lesser actions such as overtime bans – the British study explored employees' attitudes towards taking industrial action. The questions did not measure individual willingness but rather the individual's *perception* of the general willingness among employees to take industrial action in response to previous redundancies in the organization. Insecure and secure employees differed in their perceptions of the effect of the previous redundancies on the general willingness to strike – insecure employees believed more strongly that the dismissals increased the general willingness ($\chi^2 = 7.39$, $p = .025$) – but we found no difference between the two groups in their perceptions of willingness to engage in more moderate industrial action such as overtime bans.

Participants in the Israeli and Dutch studies were asked whether they would be willing to engage in industrial action to restore job security. In both countries, the number of people willing to participate in militant industrial action (strike in Israel, occupation in the Netherlands) was twice as great among employees who had experienced job insecurity as it was among relatively secure employees. In both countries, too, differences regarding moderate industrial action were less pronounced. In the Dutch study we asked employees whether they would be willing to participate in more moderate industrial action such as short work stoppages or demonstrative meetings. Four-fifths of the workers who felt insecure about their jobs were willing to do so, as against only three-fifths of those who felt relatively secure. The Israeli study also posed questions about several moderate action forms, such as meetings during working hours, slow-downs, brief work stoppages and protest demonstrations. On the whole, employees who felt insecure about their jobs were slightly more willing to participate in these moderate action forms. In both the Israeli study and the Dutch study we combined these different indicators into a measure of willingness to participate in industrial action.

The Relationship between Coping and Well-being
The analytical distinction between mental health, job satisfaction and organizational commitment as indicators of psychological well-being, on the one hand, and avoidance, individual and collective action as coping strategies, on the other, does not necessarily mean that psychological well-being and coping are mutually exclusive. To close this discussion of responses to job insecurity, we will investigate the intercorrelations of coping and well-being. Table 5.6, which presents data from the Dutch study, shows the intercorrelations among employees who feel insecure about their jobs. Since the various measures of mental health all produced similar patterns, only the correlations of Lubin's Depressive Affect Scale are exhibited. The correlations in Table 5.6 reveal some interesting patterns that deserve special attention. Perhaps not surprisingly, job satisfaction and organizational commitment are correlated. The more satisfied employees are with their jobs, the more committed they are to the organization. Depressive affect is negatively related to job satisfaction: the more depressive employees are, the less satisfied they are with their jobs. Depressive affect is not related to organizational commitment. These correlations clearly show that the three variables do not coincide completely and in fact measure separate dimensions of psychological well-being. But in their relationship to job insecurity they do coincide and may well be conceived of as constituting the psychological well-being syndrome. Among insecure employees each of the three variables is affected, that is, mental health, job satisfaction and organizational commitment are all reduced.

Table 5.6   *Pearson correlations of well-being and coping strategies among insecure workers in the Dutch study*

	1	2	3	4	5	6
Depressive affect	–					
Job satisfaction	–.29	–				
Organizational commitment	.05	.28	–			
Avoidance	.28	–.36	–.26	–		
Individual action	–.10	–.11	–.47	.11	–	
Collective action	.19	–.01	.12	–.01	.00	–

$n = 150$. When $r \geqslant 0.14$, $p < 0.05$.

Avoidance, individual action and collective action as coping strategies are virtually unrelated. In other words, job insecurity may evoke any combination of these coping strategies. Some employees may combine avoidance with job-seeking, others may engage in job-seeking and collective action, and yet others may simply withdraw and do nothing. This finding confirms data in the stress literature which demonstrate that coping strategies are not mutually exclusive (Kleber, 1982).

Perhaps the most interesting data in Table 5.6 concern the correlations between psychological well-being and coping. As it turns out, the three coping strategies have fairly distinctive patterns of correlations with mental health, job satisfaction and organizational commitment. Avoidance, for one, is evidently related to each of the three aspects of well-being. The more psychologically withdrawn employees are from their work, the less satisfied they are with their jobs, the less committed they are to the organization and the more depressed they are. Apparently, as a response to job insecurity, avoidance is part of the psychological well-being syndrome. Davy et al. (1988) report similar findings in a study of survivor responses to lay-offs.

Two active responses produce correlation patterns that are extremely interesting theoretically. Individual action shows a strong negative correlation with organizational commitment and insignificant negative correlations with job satisfaction and depressive affect: the less committed employees are to the organization, the more likely it is that they will react to job insecurity by taking individual action. This response is understandable, because to a great extent individual action consists of job-seeking behaviour. Conversely, preparedness to collective action-taking is positively related to organizational commitment: the more committed employees are to the organization, the more likely it is that they will respond to job insecurity by an increased preparedness to take collective action. Although the correlation of collective action-taking with organizational commitment is not itself significant, the difference between the correlations of individual and collective action with organizational commitment is highly significant. This pattern is exactly what one would predict on the basis of Hirschman's theory (1970) of exit, voice and loyalty: employees who are less committed to the organization will try to leave the organization to regain job security for themselves; employees who are more committed to the organization will be more likely to choose the option of industrial action to restore job security within the organization. Finally, we should note that willingness to take collective action, unlike individual action, is positively related to depressive affect. The explanation for this finding might be that collective action is more likely among those employees who are more committed to the company. Because their commitment is stronger they feel more depressed by the prospect of losing their job.

Causes, Costs and Benefits: Predicting Responses

Job insecurity, then, has many possible consequences at the individual level. It is unlikely, however, that each individual who feels insecure will manifest the same response pattern. Indeed, we found a considerable variety in the responses to job insecurity. Quite a few employees who felt insecure were hardly hurt at all psychologically; a considerable number of insecure employees refrained from any action, while many others engaged in individual or collective action or both. How come? Why do different people, all of whom feel insecure, have such different responses? In Chapter 3 we sought the answer to this question in the causal explanations of job insecurity individuals adopt, and in their perceptions of the costs and benefits connected with active responses to job insecurity. We now return to this part of our framework with empirical data. First we will describe the causal attributions our respondents made. We will then investigate how causal attributions influence the responses to job insecurity. Next, we will examine the significance of perceived costs and benefits of action-taking as determinants of coping behaviour. We will conclude this section by considering the combined impact of causal attribution and cost/benefit ratios on coping behaviour.

Attributing Causes to Job Insecurity

Situations in which job insecurity develops are often ambiguous. Consequently, among those employees who feel insecure about the future of their jobs, we can find quite a few diverging opinions about the reasons for that insecurity. Employees may have their own theories about what is happening to the company, theories that may or may not be shared by colleagues. These 'personal theories' are what interests us in this section.

Table 5.7 *Causes to which employees attribute their own job insecurity: Dutch study (n = 150: questions only asked of insecure employees)*

	Shipyard % (n = 64)	Engineering company % (n = 14)	Electronics company % (n = 72)
Individual/controllable[1]	6.1	–	2.8
Education	12.5	21.4	13.9
Work experience	1.6	–	6.9
Own effort	4.7	–	2.8
Relationship with colleagues/superior	21.9	7.1	6.9
Activities for union	9.5	7.1	5.6
Employees' efforts	40.6	21.4	27.8
Individual/uncontrollable[2]	15.4	21.4	12.5
Health	10.9	14.3	8.3
Age	53.1	64.3	48.6
Ethnic background	12.5	14.3	11.3
Social/controllable[1]	41.5	42.9	80.5
Management's decision-making	23.8	50.0	87.5
Management's efficacy in opening new markets	65.6	42.9	77.8
Concern policy	40.6	50.0	87.5
Governmental policy	95.3	–	45.8
Union policy	30.2	21.4	45.1
Social/uncontrollable[1]	60.0	78.6	73.6
Demands for products	87.5	42.9	80.6
Economic situation	89.1	78.6	66.7
New technology	46.9	78.6	63.9
Automation	26.6	78.6	49.3
Educational demands raised	47.6	21.4	73.2

[1] Percentages mentioning three or more causes.
[2] Percentages mentioning two or more causes.

In an attempt to map the causal attributions employees make when they believe that their jobs are at risk, we asked the employees in the Dutch sample to indicate, from a range of alternatives, the reasons why they were concerned about the future of their jobs. Table 5.7 presents the range of possible reasons and the percentage of insecure employees who felt that each reason applied to their situation. Employees could respond to each reason on a four-point scale ranging from 'applies definitely' to 'definitely does not apply.' The reasons are grouped in the categories we developed in Chapter 3: individual/controllable causes, individual/uncontrollable causes, social/controllable causes, social/uncontrollable causes. The four types of causal attributions are

not mutually exclusive, and generally speaking an individual's definition of the situation may encompass individual, social, controllable as
well as uncontrollable causes. For each individual, however, one type
of causal attribution may be paramount, and will colour that individual's reaction to job insecurity. To determine the importance of a
specific category of attributions for an employee, we computed scores
for each category by counting the number of reasons in a specific
category which an employee indicated were applicable in his situation.
In Table 5.7 the first row of each category presents the percentages of
employees who mentioned more than two-thirds of the reasons in that
category as applicable. We present the data for each of the three
companies separately, because some interesting differences can be
observed between companies.

The frequencies in Table 5.7 reveal striking differences between
individual and social causes. On the whole, social causes are mentioned
more frequently than individual causes, as we can see from both the
summary score and the single items. We have already discussed the
self-serving aspects of this kind of attribution (see Chapter 4). In this
context, however, we must note that attribution of job insecurity to
social causes cannot be fully explained in terms of self-serving biases.
Where the jobs of many employees are threatened because of company
decline or extensive reorganization, it is only natural to find many
workers who blame social causes for their job insecurity.

This phenomenon is also reflected in the different explanations
provided by the employees of each of the three Dutch companies. In
the eyes of the employees of the shipyard, there is no mistaking the
problem: the general economic situation and decline in shipbuilding,
together with an unfavourable governmental policy towards their
shipyard, are the main reasons for concern about the future of their
jobs. To the employees of the electronics company, on the other hand,
it is quite clear that poor managerial decision-making both at the plant
and in the company headquarters is at fault. Employees of the
engineering company, on the other hand, were generally less certain
about the origins of their situation. For them, none of the possible
reasons stands out as especially important. Clearly, in both the
shipyard and the electronics company, employees agreed on some
elements of a shared definition of the situation. Among the employees
of the engineering company we did not find such a consensus.

In face of the obvious externality of some of the threats, it is amaz-
ing that so many employees nevertheless attribute their insecurity to
individual causes, especially age and education. Apparently, quite a
few employees fear that the burden of lay-offs (or whatever measures
management takes) will not be distributed evenly or objectively, that
is, irrespective of personal characteristics such as education, work
experience, efforts, union involvement, health, age or ethnic

background. Whether these fears are realistic is not at issue here. At the very least management has not been able to prevent them from developing. Moreover and more important, these fears influence an individual's reaction to job insecurity (as we will see in the next sections).

Looking at the separate individual factors, we find an impressive percentage of the Dutch employees referred to their age as a reason for concern. At least in the Dutch situation, age is apparently seen as a disadvantage, unless one is young. The decisive age is about forty; employees who are younger than forty do not refer to their age as a reason for concern. Smaller yet still impressive is the number of people expressing concern about the future of their jobs because of their education or their relationships with colleagues or superiors. As for employees who cite their ethnic background as a cause for concern, the percentage appears to be moderate – but only at first glance. We must remember that only a small percentage of the workforce in the three companies comes from ethnic minorities. Indeed, of those employees who belong to an ethnic minority, almost all mention ethnic background as a reason for concern.

If we examine our data from the perspective of controllable versus uncontrollable causes, we see that many employees who feel insecure about their jobs refer to uncontrollable causes – social or individual – as the reason for their concern. This attribution does not necessarily mean that these people see their situation as completely uncontrollable, since one person may very well cite both controllable and uncontrollable causes. Nevertheless, it tells us something about the way in which employees who fear for their jobs define their situation, and it indicates the kinds of opportunities they might look for to improve that situation.

The Sources of Causal Attributions

Differences in causal attribution may derive from objective, individual and social sources. To some extent we have already discussed the objective origins indirectly in the previous section where we compared the employees from the three Dutch companies. Objectively, the situations of the three companies were rather dissimilar, as our description in Chapter 1 suggests. The causal attributions presented in Table 5.7 reflect these differences in the objective situations. Objectively observable characteristics of the situation, such as the decline in shipbuilding together with governmental policy in the case of the shipyard or poor management in the case of the electronics company definitely influenced employees' causal attributions. In the engineering company, where it was more difficult to identify the objective causes of decline we found less of a shared definition of the situation.

On the other hand, individuals may be more or less predisposed to

Table 5.8 *Percentages of employees receiving information from specific sources: Dutch study* (n = *311*)

	Shipyard % (n = 106)	Engineering company % (n = 75)	Electronics company % (n = 130)
Company management	39.6	56.0	54.6
Personnel magazine	60.4	54.7	50.0
Bulletin board	77.4	65.3	83.1
Superior	50.9	59.5	70.0
Minutes of workers' council	72.6	65.3	46.2
Members of workers' council	40.6	32.0	22.3
Union representatives	27.4	22.7	10.8
Union newspaper	34.0	16.0	6.2
Colleagues	67.0	64.0	83.8
Mass media	78.3	44.0	86.2
Friends/acquaintances	16.0	13.3	20.8

Respondents could mention more than one source.

define their situation in a specific way. Individuals can be distinguished according to their predisposition to base their attributions on either internal or external loci of control (Rotter et al., 1972). We can assume that people who in general see their lives as determined by external uncontrollable causes will see the source of their job insecurity in the same way. The Dutch study revealed exactly this correlation: the more 'external' employees are – that is, the stronger their belief that their lives in general are determined by external uncontrollable forces – the more likely they are to attribute their feelings of job insecurity to uncontrollable causes, both social and individual (Pearson correlation .34).

The social origins of causal attribution lie in the fact that people usually do not make up their minds in isolation. The meaning of an event is socially constructed through the interaction of all those involved in the same situation using information disseminated by sources such as management, workers' councils, unions or mass media. Table 5.8 suggests the relative significance of various sources of information.

First, it is interesting to see the relative significance of colleagues as sources of information, especially in the electronics company. This statistic underscores the importance of social interaction with people in one's immediate environment in assessing the meaning of what is going on. Second, company sources such as management, the personnel

Table 5.9   *Pearson correlations of sources of information and causal attributions: Dutch study*

	Management	Superior	Members workers' council	Union officials	Colleagues	Mass media
Individual/ controllable	.15	.06	−.10	−.17	−.04	.12
Individual/ uncontrollable	.13	.25	.08	.00	.08	.18
Social/ controllable	.09	−.01	.21	.20	−.08	−.13
Social/ uncontrollable	.00	.02	.15	−.04	−.13	−.14

$n = 150$; when $r \geqslant .14$, $p < .05$; differences between correlations $\geqslant .14$ are significant at $p < .05$.

magazine, the bulletin board and one's superior are significant sources of information about the company's situation. Unions are relatively insignificant compared to other sources of information. Even in the case of the shipyard, where the unions have a very strong position, little more than one-third of the employees referred to unions as a source of information. We must point out, however, that the modest role of the unions is to some extent compensated for by the more significant role of the workers' council, especially the reports from the council meetings which are distributed inside the company. Unions have central positions in the workers' councils, and we can assume that part of their message is transmitted through the channels of the council. Mass media were important in the shipyard and in the electronics company. Friends and acquaintances were the least significant sources among all three groups of employees. In all likelihood the main sources of information – company sources, workers' councils and colleagues – work simultaneously. That is, company sources and councils presumably provide the input for discussions among groups of colleagues.

Insight into those sources of information which employees depend on would be of little use if an individual's understanding of his or her situation were not influenced by those sources. Table 5.9 investigates the relationship between the employees' choice of causal explanation and what they report to be their primary sources of information. The table presents Pearson correlations between the number of causes in a specific category which the employees mentioned and a specific source of information. Examining these correlations, we find some interesting patterns. The greater the employees' dependence on company sources, the more inclined they were to attribute their feelings of job insecurity to individual causes, that is, to controllable or uncontrollable personal characteristics. Correlations with the unions and the workers' councils produce precisely the opposite pattern: the more employees relied on union sources or members of the workers' council, the more they attributed their feelings of job insecurity to social causes. Further, they were less likely to make attributions to individual/controllable causes. Mass media seem to produce the same effect as company sources: they too made individual attributions more likely and social attributions less likely. Perhaps because of the influence of these counteracting sources, interactions with colleagues did not make one pattern of causal attributions more likely than another.

Although we must not be too hasty in drawing conclusions from this evidence, we can offer at least one possible explanation for these patterns: it is likely that company sources communicated, implicitly or explicitly, the weight given to personal characteristics in decisions about job losses. On the other hand, sources on the side of the union and the workers' council seem to have emphasized that controllable/ social factors, especially managerial decisions, are responsible for the uncertainty. Employees who rely on these sources, then, are less likely to consider personal characteristics as factors determining job losses.

To summarize, the differences in causal attribution we found were related to (1) the different situations of the companies, (2) the different personality characteristics of the employees affected and (3) the employees' different sources of information. Although we did not fully demonstrate the social construction of meaning, our findings imply that interaction among colleagues is an important element in the process of defining the company's situation.

Causal Attribution and Responses to Job Insecurity

Do causal attributions affect responses to job insecurity? Do the statistics bear out our hypothesis that employees who attribute their feelings of job insecurity to personal characteristics are more depressed than employees who blame social factors for their insecure situation? Are employees who blame controllable causes more likely to respond actively than employees who blame uncontrollable causes? The correlations presented in Table 5.10 can help answer these questions. These correlations, which present data for Dutch employees who feel insecure about the future of their jobs, indicate the extent to which psychological well-being and coping strategies are related to the type of causal attributions an individual makes.

The correlations between causal attribution and psychological well-being are clearer than those between causal attribution and coping behaviour. To begin with well-being: attributing one's job insecurity to individual characteristics, whether controllable or uncontrollable, further reduces job satisfaction. More specifically, attributions of job insecurity to individual/*un*controllable causes reinforces symptoms of

Table 5.10   *Pearson correlations between causal attributions and responses to job insecurity: Dutch study*

	Depressive affect	Job satisfaction	Organizational commitment	Avoidance	Individual action	Collective action
Individual/ controllable	.11	−.38	−.11	.14	.11	.20
Individual/ uncontrollable	.32	−.36	.13	.25	−.31	.14
Social/ controllable	.05	−.06	−.30	.14	.16	.20
Social/ uncontrollable	.06	−.05	−.02	.14	−.02	.06

$n = 150$; when $r \geqslant .14$, $p < .05$; differences between correlations $\geqslant .14$ are significant at $p < .05$.

diminished mental health: if employees attribute their insecurity to uncontrollable characteristics such as age, health or ethnic background, their mental health is negatively affected. Interestingly, attribution to social but controllable causes is related to reduced organizational commitment. In other words, attributing job insecurity to poor management – the main element of this causal explanation – reinforces the negative impact of job insecurity on commitment to the company. Attribution to social but uncontrollable causes does not affect the relationship between job insecurity and psychological well-being.

As for coping strategies, we had hypothesized that attribution to individual/controllable causes makes individual action more likely, that attribution to individual/uncontrollable causes produces avoidance, that attribution to social/controllable causes encourages collective action and that attribution to social/uncontrollable causes leads to avoidance.

The data in Table 5.10 confirm some of the hypothesized relationships between causal attribution and coping behaviour but fail to confirm others. Contrary to our expectations, individual/controllable and social/controllable attributions make all of the three coping strategies more likely. That is, attributions to controllable causes in general foster avoidance, individual action and collective action equally. In line with our expectations, attribution to individual/uncontrollable causes does indeed make individual action less likely and avoidance more likely. Since age, health and ethnic background are the major components of this type of causal attribution, and job-seeking is the primary form of individual action available, this negative correlation underscores the difficulties that some job-seeking employees anticipate in finding another job. (We will return to this issue in our next section.) One unexpected finding was that individual/uncontrollable attributions increase the likelihood that employees will pursue collective action. This positive correlation seems to indicate that these employees see in collective action a possible means of protecting themselves against biased lay-off policies. Finally, as we expected, social/uncontrollable attribution makes avoidance more likely.

We can conclude, then, that causal attributions do make a difference in the impact job insecurity has on psychological well-being and on the adoption of coping strategies. Attribution to individual causes – especially those which are uncontrollable – reinforces the negative impact of job insecurity on psychological well-being. For obvious reasons, attribution to social/controllable causes reinforces the negative impact of job insecurity on commitment to the company. Coping behaviour seems less affected by causal attributions. The only distinct patterns that emerge are those related to the two explanations in terms of uncontrollable causes: attribution to individual/uncontrollable causes makes avoidance more likely, individual action less likely and collective action more likely; attribution to social/uncontrollable causes makes avoidance more likely and is unrelated to active coping strategies. On the other hand, attribution to controllable causes makes all of the three coping styles more likely. Apparently, if one attributes one's job insecurity to controllable causes, both kinds of active responses – individual or collective – seem reasonable. The hypothesized distinction between individual and collective action did not materialize. The positive correlation with avoidance may be related to the costs of action-taking, since avoidance was thought to be the most likely response when the costs of taking action are too high for the individual. To determine why employees who explain their job insecurity in the same terms (that is, controllable causes) should resort to three different strategies, we must take other factors into account.

The Costs and Benefits of an Active Response
Active responses aim at restoring job security, either individually, by changing one's personal situation, or collectively, by changing the organizational circumstances. In our previous section we saw that, while causal attributions certainly influence the adoption of coping strategies, they do not fully explain the differences in coping behaviour among employees. The costs and benefits of active responses must be taken into account as well. In examining these determinants of coping behaviour, we will consider individual action first and then collective action.

Individual Action The major benefit of individual action is, of course, the acquisition of a more secure job. The more important it is for you to have work, the more you will benefit from obtaining a more secure job. But who guarantees that you will find one? What are, after all, your chances in the labour market? These questions present themselves to anyone who is about to undertake the search for a more secure job. Individual action, then, is in our view stimulated by the importance of having work on the one hand and by one's chances in the labour market on the other.

The Dutch study gave us the opportunity to test this hypothesis. We assessed both how important it was for the respondents to have work and how hopeful they were about their chances in the labour market. Although all the respondents affirmed the importance of having work, the degree of importance varied from individual to individual. We further asked the respondents to estimate their position in the labour market and to indicate whether they thought it would be difficult to find another job. We found that insecure employees engaged in individual action to restore job security more frequently the more important it was for them to have work (Pearson correlation .18) and the more optimistic they were about the opportunities for finding another job (Pearson correlation .34 with estimated position in the labour market; Pearson correlation −.41 with the belief that it is going to be difficult to find another job). Thus, we can conclude that insofar as the importance of having work and the chances of finding a more secure job stimulate individual action, that action is indeed influenced by the perceived benefits associated with it.

Collective Action Collective action in response to job insecurity may be similarly influenced by the perceived costs and benefits of action-taking. To undertake collective action to restore job security, an individual expects that collective action will succeed in restoring job security and that the balance of costs and benefits of participation will not be too unfavourable. The general category 'expectations of success' can be broken down into several elements: expectations about one's own contribution to achieving success, expectations about the behaviour of one's colleagues and expectations about the successful outcome of collective action. Our present research is restricted to the last two expectations, that is, the expectations about the behaviour of colleagues and expectations about a successful outcome. We defined the costs and benefits of participation in terms of the expected positive or negative reactions of significant others such as one's colleagues, family and friends, and in the perceived risks for one's position in the company if the individual were to take part in industrial action.

Table 5.11, based on data from the Dutch sample of employees who felt insecure about their jobs, presents the correlations of these variables with willingness to participate in collective action. Generally speaking, employees who felt insecure about their jobs were willing to take part in collective action to secure their jobs, if they expected that collective action would have success, and if the costs and benefits of

Table 5.11 *Pearson correlation of the costs and benefits of action-taking and willingness to take part in collective action: Dutch study*

	Willingness to undertake collective action
Expected participation of colleagues	.42
Expected success	.20
Expected reaction of family, friends and colleagues	.46
Perceived risk for one's position in the company	.26

$n = 150$. When $r \geqslant 0.14$, $p < 0.05$.

participation appeared favourable. Looking at each of these two conditions individually, we found that, in the case of success expectations, those employees who were more confident about the participation of their colleagues – and to a lesser extent those employees who were more convinced that collective action would be effective – were willing to participate. In the case of perceived costs and benefits, we found that damage to one's position in the company – and, in particular, negative reactions of one's friends, colleagues and family – made employees less willing to participate in collective action. Together, these determinants were able to explain 35 percent of the variance in willingness to participate in collective action among employees who felt insecure about their jobs. And of these four determinants, expectations about the participation of one's colleagues and the expected reactions of significant others proved the most important.

Obviously, the perceived costs and benefits of collective action are important determinants of willingness to take action. These findings confirm our general model of participation in collective action, but one questions remains: is the experience of job insecurity in itself a factor in the explanation of an employee's willingness to engage in collective action to restore job security? To arrive at an answer we conducted a regression analysis among both the secure and the insecure employees in the Dutch sample. Our purpose was to assess the combined impact of job insecurity, success expectations and perceived costs of participation on employees' willingness to participate in collective action.

Two of the affective responses discussed earlier, anger and anxiety, were also included in the analysis, because we wanted to investigate the impact of these emotional reactions to job insecurity on willingness to take part in collective action. We chose anger and anxiety because they were the two responses that were significantly correlated to collective action-taking. A look at Table 5.12 reveals that job insecurity significantly increases employees' willingness to participate in

Table 5.12   *Multiple regression of willingness to participate in collective action on job insecurity, and costs and benefits of participation: Dutch study*

	Standardized beta
(a) Job insecurity	.12[1]
(b) Anger	.20[2]
(c) Anxiety	−.14[1]
(d) Expected participation of colleagues	.31[3]
(e) Expected success	.10[1]
(f) Expected reaction of family, friends and colleagues	.29[3]
(g) Perceived risk for one's position in the company	.11[1]
r^2	.38

$n = 308$.
[1] $p < .05$.
[2] $p < .01$.
[3] $p < .001$.

collective action independent of the other determinants. As our previous analyses suggested, success expectations and favourable cost/benefit balances considerably increase one's willingness to participate. Perhaps the most interesting finding concerns the two affective responses: anger increases willingness to participate, while anxiety reduces willingness. In evaluating the significance of this finding, we must recollect that both anger and anxiety were positively related to job insecurity. These results indicate that job insecurity makes employees more inclined to take part in industrial action, and that this inclination is stronger or weaker, depending on the affective response job insecurity evokes. Employees whose primary response to job insecurity is anger are particularly willing to take collective action. Employees who respond with anxiety as well, however, are less willing to take part in collective action.

Testing the Model
Our model must do more than demonstrate that causal attribution and cost/benefit ratios influence coping behaviour, for the model stipulated not only that causal attribution makes a particular coping strategy more pertinent but that the balance of perceived costs and benefits determines the relative appeal of the several coping strategies. Consequently, we must demonstrate that both causal attributions *and* perceived costs and benefits of action-taking are needed to explain coping behaviour. Moreover, we must demonstrate the validity of our assumption that individuals who perceive the cost/benefit ratio of active responses as too unfavourable will resort to avoidance. Table 5.13 presents the results of three regression analyses among insecure employees from the

Table 5.13 *Causal attributions and perceived costs and benefits as predictors of coping strategies: multiple regressions from the Dutch study*

	Beta
Avoidance	
Causal attributions	
Individual/uncontrollable	.25[2]
Perceived costs and benefits active responses	
Expected success collective action	−.20[2]
r^2	.13
Individual action	
Causal attributions	
Individual/uncontrollable	−.18[1]
Individual/controllable	.19[1]
Perceived costs and benefits of individual action	
Expected problems in finding a more secure job	−.36[3]
Importance of having work	.21[2]
r^2	.26
Collective action	
Causal attributions	
Individual/uncontrollable	.13[1]
Social/controllable	.23[2]
Perceived costs and benefits of collective action	
Expected participation of colleagues	.30[3]
Expected success	.18[2]
Expected reactions of significant others	.32[3]
Perceived risk for position in comapny	.13[1]
r^2	.43

$n - 130$.
[1] $p < .05$.
[2] $p < .01$.
[3] $p < .001$.

Dutch study. In these analyses causal attributions and perceived costs and benefits of action-taking were related to avoidance, individual action and collective action. From these analyses we may conclude that *both* causal attributions *and* cost/benefit considerations are needed to explain the adoption of specific coping strategies, although the amount of variance explained differs for each of the three coping strategies. We were most successful in explaining willingness to act collectively: 43 percent of the variance could be explained by causal attribution and the perceived costs and benefits of collective action. According to these results, we can expect employees who feel insecure about the future of their jobs to be more willing to undertake collective action to restore job security:

1 the more they attribute their insecurity to individual, uncontrollable causes and/or social, controllable causes,
2 the more convinced they are that their colleagues will also engage in collective action, and the greater their expectation that the collective action will succeed,
3 the stronger their belief that significant others will support their action-taking and that their position in the company would not be jeopardized by their action-taking.

For individual action, we were able to explain 26 percent of the variance in action-taking. Employees who felt insecure were more willing to engage in individual action:

1 the less they attributed their insecurity to individual, uncontrollable causes, and the more to individual, controllable causes,
2 the more problems they anticipated both in finding a more secure job and in changing from one job to another, and the more important it is to them to have work.

We were least successful in explaining avoidance: we could account for only 13 percent of the variance of avoidance. Employees who felt insecure were more inclined to resort to avoidance:

1 the more they attributed their insecurity to individual, uncontrollable causes,
2 the lower their expectations that collective action would succeed.

Avoidance, that is, psychological withdrawal from one's work as a response to job insecurity, appears to be unrelated to the perceived likelihood of finding another job; the Pearson correlation between avoidance and the perceived labour market position is virtually zero. This result contradicts our assumption that people will resort to avoidance if the costs/benefit ratio of individual action-taking is unfavourable.

Attribution to individual, uncontrollable causes influences each of the three coping strategies. It makes employees more likely to withdraw, it makes them less active individually, but it also makes them more active collectively in response to job insecurity. Attribution to controllable causes makes it more likely that employees will respond actively. Significantly, however, attribution to controllable, *individual* causes makes individual action more likely, whereas attribution to controllable, *social* causes makes collective action more likely. This finding is exactly as we predicted.

Summary

For many people job insecurity appears to have a detrimental effect on their psychological well-being. Job insecurity is related to impaired mental health, reduced job satisfaction and reduced organizational commitment. Since our data were basically correlational, we must be careful in drawing conclusions about causal relations. But one thing is clear: job insecurity is not a healthy state of mind. There is little reason to believe that it has any positive consequences for the affected person.

Its behavioural consequences reveal that job insecurity has few positive implications for the organization either. Indeed, if insecure employees respond actively, their activity consists of attempts to restore job security by seeking another job or by engaging in industrial action. In this context it is interesting to note that organizational commitment serves as an intervening variable: the less committed employees are to the organization, the more likely they are to try and leave it; the more committed they are, the more likely it is that they will engage in industrial action. Perhaps paradoxically, the employees who would most prefer to stay if the company were capable of providing them with greater job security are those who are most likely to engage in industrial action. Since industrial action is often directed against company management, however, management may be reluctant to negotiate with or even listen to the activists. On the other hand, calmness within the workforce can easily be misinterpreted by management as a sign that there is nothing to worry about. But calmness might well be a sign of avoidance instead: the employees might be psychologically divesting themselves of their commitment in the face of a situation that could end with their job loss. As a result of reaction, the company may find itself in a downward spiral. Part of the workforce is reluctant to invest psychologically in the company any longer, either because it is trying to leave and/or because it is withdrawing. Part of the workforce, while remaining committed to the company, is predisposed to engage in industrial action, but is ignored.

Our findings on causal attribution and cost/benefit considerations as determinants of coping behaviour may provide management with some leverage. As we have demonstrated, both factors play an important role in determining the way employees cope with job insecurity. Causal attribution to account for one's job insecurity may either intensify or moderate its detrimental effects. Attribution to individual characteristics diminishes psychological well-being. If these characteristics are considered uncontrollable, the chances are that the affected employees will withdraw psychologically, thus ultimately reducing the company's productivity. Employees who consider the individual characteristics to which they attribute their job insecurity as controllable will be less likely to opt for avoidance, but inclined to leave the company. Clearly, then, it is in the company's interest to counteract employee self-imputation. Blaming the workforce for the company's troubles is counter-productive since it may encourage turnover and reinforces psychological withdrawal. Blaming social forces beyond anyone's control may also be counter-productive, because it, too, increases the likelihood that employees will adopt avoidance as their coping strategy. If possible, a firm should define its troubles such that the difficulties are due to social but controllable forces. In that case both management and workforce are provided with specific points on which to focus their efforts. (The shipyard is an interesting case in point.) Nor does it seem to be a bad idea for management, if the fault is (partly) theirs, to admit that such is the case. At the same time, management must not take a negative view on industrial action. After all, for the the most part such action is taken by employees who would rather see improvement at their place of work than withdraw their commitment. If management increases the cost of industrial action by imposing sanctions or by blocking any chance of success, these activist employees too may prefer to leave or withdraw psychologically.

Referring our findings back to the framework developed in Chapter 3, we may conclude that the model has been useful in analysing individual responses to job insecurity. Job insecurity appears to have the predicted detrimental effect on individuals; each of the coping strategies we distinguished did occur; and both causal attributions and perceived costs and benefits of active responses were needed to explain the adoption of coping strategies, though we could explain collective action better than individual action and avoidance.

Our research on the impact of job insecurity at the individual level is related to three different literatures: stress literature, attribution literature and collective-action literature. To close this discussion of individual responses to job insecurity, we will offer some conclusions relevant to these literatures.

In our approach to job insecurity we have expanded the *stress framework* with the concept of attribution and the notion of perceived costs and benefits of active responses. Moreover, rather than restricting active responses to individual action we broadened this category to include collective action as a possible response to stress. Many people, depending on their definition of the situation, found collective action – more specifically industrial action – a feasible response as well.

Taking the well-known categorizations of *causal attributions,* we have expanded the notion of controllability by including controllability through collective action. On many occasions, circumstances that cannot be controlled by the individual may be controlled collectively – as history demonstrates. Perhaps not surprisingly, beliefs about how well a situation can be controlled through industrial action turn out to be especially important in generating the willingness to participate in collective action.

This discussion of the way attribution theory contributes to the understanding of real-life situations requires one further remark. The attribution literature contains several categorizations of causal explanations: in terms of individual versus social, controllable versus uncontrollable or stable versus unstable. As we found out, it is not always easy to apply the neat categories of attributions theory to real life (cf. Hewstone and Jaspars, 1984; Schaufeli, 1988).

Some of our findings are relevant to *collective action theory.* First and foremost, when job insecurity was controlled for success expectations and selective incentives, it did increase employees' willingness to participate in industrial action. This result is pertinent to the ongoing debate in the literature on collective behaviour and social movements about the significance of discontent for collective action. Contrary to the arguments frequently presented in the literature, our findings suggest that discontent *is* important in motivating individuals to engage in collective action.

In the light of the dominance of rational models in the collective-action literature, our findings about the role of emotional reactions such as fear and anger in explaining people's willingness to take part in industrial action are especially revealing. The opposite effects of fear and anger on individual decisions to take collective action help explain the contradictory findings reported in the literature about the impact of a deteriorating situation on the willingness of individuals to protest. Although one event may be experienced by many people, different individuals will respond in different ways. Some may become frightened and less willing to protest, while others will become angry and more eager to take action. Our findings concerning causal attribution and collective action are equally important. A considerable proportion of the variance in willingness to engage in industrial action could be accounted for by causal attributions.

6

Industrial Relations and Job Insecurity: A Social Psychological Framework

Jean Hartley

We pursue issues of job insecurity and industrial relations in more detail in this and the following chapter. Here we look at individual and collective behaviour at workplace level, investigating particularly how job insecurity may affect trade union organization and union–management relations. There has been some speculation about the effect of fears about job loss on industrial relations in recent years, fuelled particularly in the decade of the 1980s when recession and restructuring dominated much industrial-relations commentary. Such discussion is also relevant to the 1990s with restructuring and change continuing apace. Until recently, the analysis of job insecurity has been unsatisfactory, as we shall demonstrate. This chapter aims to show how a social psychological framework is essential to understanding some of the changes which have been taking place in industrial relations. The following chapter then uses case-study material to illustrate and investigate aspects of the framework. In this way, these two chapters can contribute to the debates about restructuring and industrial relations. We believe the material also goes further than this to explain aspects of union–management relations in situations of rapid organizational change. The chapters are based on British debate and evidence primarily. However, it will be argued that they have a wider relevance.

In arguing for a social psychological dimension in analysing change and industrial relations, the intention is to supplement the analyses which are derived from other levels of analysis, notably the macro-economic and institutional levels. A psychological approach to industrial relations will always be insufficient on its own (Hartley and Kelly, 1986; Hartley, 1988). Equally, however, a macro-economic or institutional framework standing alone lacks the analyses of individual and collective perceptions and motivations which are needed to explain industrial-relations processes and outcomes. Indeed, it is argued that the failure to understand some of the changes in industrial relations over the last decade stems from a poor understanding of perceptions, attitudes and the bases of behaviour (see also Kelly and Kelly, in press). So there is a need to draw on several levels of analysis to understand the broad social forces which create constraints on individuals and collectivities (Hartley, 1988).

The concept of job insecurity is central to our analysis. Although the industrial-relations literature has often assumed that job insecurity may be important this has rarely been made explicit. For example, as economic recession and industrial and technological restructuring began to bite across the world in the early 1980s it was felt by many commentators that fears about unemployment would affect workers' trade union involvement and especially their ability and motivation to resist managerial power. Although the term job insecurity itself was only rarely used it was often implicit in ideas that workers were concerned about losing their jobs and were afraid of becoming unemployed. There were several calls for research to elucidate how industrial relations was being affected by high levels of unemployment and redundancy (Towers, 1982; Bright et al., 1983; Hunter, 1980; Lane, 1982; Brown, 1983a, 1986; Strauss, 1984; Winchester, 1983).

The assumption that job insecurity affects industrial-relations behaviour is pervasive and lodged at several levels. At a macro-economic level, the argument is well rehearsed: recession results in higher levels of unemployment, which in turn leads to moderated wage demands, increased productivity and greater co-operation between workers and management. In this way recession ultimately has a beneficial effect on the economy since it drives change, making organizations leaner and fitter and encouraging more flexible attitudes and behaviours on the part of employees. Some commentators explicitly mention 'the fear factor' as fuelling the changes. For example, Metcalf states: 'The crucial factor explaining the growth of labour productivity . . . was the employment reduction . . . experienced between 1980 and 1982. Fear must be what matters here. First, there was the fear of bankruptcy on the employers' side . . . Second, employees were fearful for their jobs and prepared to work harder' (1989: 19). However, fear is asserted, not examined. It is a post-hoc variable to explain management and worker behaviour changes. It has even been assumed, though more contentiously, that the changed economic conditions have resulted in 'new realism' – changed attitudes on the part of workers, occasioned by their fear of job loss. The mechanism of fear is essential to this view, although often implicit (for example, Bassett, 1986; Gennard, 1985; Hawkins, 1985; Edwards and Heery, 1989).

In practice, the relationship between the national level of unemployment and industrial-relations behaviour has not been so straightforward (see Martin, 1987; Hunter, 1980) and modifications have been put forward, for example that it is the rate of increase in unemployment rather than its absolute level which acts to curb worker militancy and wage demands. This still implies a concept of fear of unemployment. There are significant issues still to be resolved here, for example whether the macro-economic factors are believed to operate through perceptions of national or local unemployment, through unemployment or job losses (such as the rate of redundancies) or other factors. Although a crude mechanism of economic force and industrial-relations behaviour has been promoted, the processes remain obscure because there has been no attention paid to worker perceptions or their sense of job insecurity. From a psychological point of view, the assumption cannot be made that objective conditions map on to subjective reality.

At the institutional level of analysis, recession and restructuring was initially interpreted as having shifted the balance of power away from workers and trade unions towards the employer. The weakening of union power was attributed, either implicitly or explicitly, to the fear of unemployment (for example, Edmonds, 1984; Towers, 1982; Ogden, 1981; Bright et al., 1983; Kelly, 1984).

It was believed that the changed balance of power would be reflected in modified industrial-relations structures and procedures. Initially commentators had proposed that high unemployment would lead to the reassertion of managerial power (even macho management), with unions repressed, destroyed or sidelined. This outcome would be reflected in weaker union organization, reduced union recognition and collective bargaining, poorer wages and employment conditions, and a lower level of industrial conflict. The analyses early in the 1980s at both the macro-economic and institutional level could not determine either what the impact of changed economic conditions was, nor what explained the changes which did take place (see Terry, 1986, and Legge, 1988, for summaries). For both of these questions it is necessary to have an understanding of processes taking place at the level of the workplace: how wider social forces were being interpreted and how industrial-relations actors perceived the causes and consequences of economic change. In the initial assumption of weakened trade unions, job insecurity was an implicit part of the mechanism, though so taken for granted that its role as an intervening variable between economic change and outcome was not explored. For example, 'A number of well-publicised examples have shown how difficult it is for unions to resist closures and redundancy, demanning and speed-up, when faced with the market forces of the recession' (Spencer, 1985: 22). Another example is Hyman who talks of weakening trade union power because of 'growing surpluses in the labour market' (1987: 117). The article by Gennard (1985) describes the impact of recession on union membership without stepping any closer to job insecurity than saying that 'new realism' may be developing on the part of unions. Or take Price and Bain, for whom it is sufficient to point to the broad impact of unemployment without being specific: 'Unemployment appears to be levelling off but it is likely to remain at present levels at least until the end of the decade and to continue to undermine union membership and union strength' (1983: 158). Yet they do not suggest any mechanisms, whether at collective or individual level, to explain why a relationship between unemployment levels and union strength is expected. Terry comes closest to explaining this relationship although he then veers away again. Pondering possible reasons for the impact of unemployment on industrial relations and especially union behaviour, he suggests:

> A more persuasive argument comes from the view that workers' morale has been sapped by evidence of redundancy and closure: that unemployment works through affecting bargaining strength by posing eventual unemployment as a stark and credible consequence of workers' actions. *But although this provides us with an intervening link between unemployment and union weakness, it does not get us any closer to a satisfactory institutional account of events.* (1986: 176-7, my emphasis)

It is surprising that a 'persuasive' intervening link is so quickly dismissed. An analysis needs to draw on concepts, from whatever quarter, where they help to clarify reasons and processes underlying institutional forms and trends. It is even more surprising that the institutional literature generally has been so lacking in curiosity about the mechanisms underlying the presumption of a relationship between unemployment and the behaviour of those remaining in work.

Yet curiosity is essential if we are ever to get towards any detailed understanding of worker and union response to recession and restructuring. We need a clearer picture of the force called 'unemployment' and how it operates on those still in work, rather than an ambiguous concept which is used in a generalized, post-hoc way. We need research which looks in detail at some of the presumed forces acting on workers, unions and managements in order to tease out the inter-relationships between a variety of social, economic and political forces.

Without the concept of job insecurity, we can see the poverty of a solely institutional approach to industrial relations, which elevates a concern with the structures of industrial relations over the *behaviour* and *social processes* of the actors, both individuals and groups, which make up these institutions. Although an institutional approach has been very valuable in charting major trends and in pointing to wider economic and political forces, its limitations become evident in assessing change in workplace industrial relations.

Although some writers have made assumptions about fear of redundancy or the timidity of unions in recessionary conditions, this has not been investigated and indeed, as we have seen, is viewed as largely peripheral to an institutional approach. Yet we need to understand how workers perceive their circumstances and what motivates them to act or not if we are to understand restructuring and industrial relations. For example, union power may be partly related to objective features such as membership levels and density as Price and Bain (1983) suggest, but it is also related to the extent to which ordinary members are willing to take part in union activities. This requires an examination of their beliefs, values and attitudes, and the ways in which these beliefs can be shaped by economic and institutional changes and mobilized by union activists (see, for example, Klandermans, 1984a, 1986; Batstone et al., 1978). Part of the structure of beliefs and values in a time of major change, uncertainty and possible redundancy may concern views and feelings about job insecurity.

The lack of concepts rigorously applied at the level of individual or collective behaviour or processes has left the institutional approach relatively unprepared theoretically for the impact of recession and restructuring. Although it has been possible to say that 'the effect of high unemployment has been undeniable' (Brown, 1986: 161), researchers have said little about the mechanisms underlying those effects.

So far, we have argued that the macro-economic and institutional approaches to understanding industrial relations in recession and restructuring are flawed to the extent that they ignore the concept of job insecurity despite sometimes making implicit use of that or similar terms. It is now time to build an alternative framework to explain better the relationship between industrial relations in conditions of restructuring.

Such a framework places job insecurity centre stage – not because it is more important than other variables which have been investigated so far in the industrial-relations literature but because it provides an important link between the macro-economic factors surrounding unemployment, institutional responses and group and individual behaviour in the workplace.

A Social Psychological Model of Industrial Relations

In this section, we develop a set of propositions about job insecurity as an intervening variable between economic conditions and industrial-relations processes and outcomes. These propositions will then be tested with data in the case study in the following chapter.

First, is job insecurity a useful concept and to what is it related? A social psychological approach needs to investigate job insecurity rather than assume it exists. Second, we shall look at the organizational antecedents of job insecurity. While previous chapters have examined the individual-level antecedents of insecurity, here we look at the relationship between organizational decline and insecurity. Then, we look at the consequences of job insecurity for industrial relations. These consequences are of two types. So the third area to investigate is how job insecurity relates to individual-level variables of industrial relations, such as attitudes, beliefs and behaviour. For example, does job insecurity make workers more or less inclined to join or participate in trade union activities in the workplace? Does it make an employee more or less motivated to be a workgroup representative? Does job insecurity affect willingness to take industrial action? The fourth area to investigate in this approach is also a consequence of job insecurity. These are the collective processes and outcomes which occur where job insecurity exists. So we investigate the impact of job insecurity on union organization. Does insecurity affect the stability and tenure of representatives? How does it influence relations between representatives, and between them and their members? Casting this in psychological terms, what is the impact of job insecurity on intra-group processes?

Finally, the approach explores union–management relations. Much of the institutional literature has suggested that union–management relations have been modified by economic change. It has been suggested that there have been some changes in attitudes between parties: discussions of 'new realism' or 'macho management' both contain assumptions not only about changes in behaviour but also about attitudes towards the other party. The current approach draws on the social psychology of inter-group processes to help explain union–management relations. In turning to union–management behaviour, the power of each party is examined.

Job Insecurity
The concept of job insecurity should by now be familiar (see especially Chapter 2). However, it is worth briefly recapping on what we understand by the term because it is the central concept in this and the following chapter and because it is a linking concept between decline and industrial relations. Job insecurity, following earlier definition (see Chapter 2), is taken to be 'the concern about the future of one's job'. It is a global concept of feelings of uncertainty about whether the job will continue or not. Operationally, job insecurity is conceived as being expressions of worry, concern and probability of the loss of one's own job occurring within the next year. It has both affective and cognitive components. Job insecurity is based on subjective assessment by the individual of the risk of their own future job loss. In this research, it has been operationalized as a three-item scale.

Organizational Decline
While attitudes and perceptions are important in understanding industrial relations, it is also essential to consider the organizational constraints within which much collective behaviour takes place. The approach here emphasizes the organizational constraints imposed by decline. The consideration of job insecurity needs to be within the context of such organizational processes. Decline may affect how

workers perceive organizational events and also may have a direct effect on the resources available, especially to management, to conduct industrial relations. Decline is the starting point for considering how job insecurity and industrial relations are related. Although job insecurity can occur in a variety of settings including organizational growth (Borg and Hartley, 1989), we have examined its occurrence in a setting of organizational decline for several reasons. First, because organizational downturn is the dominant context within which job insecurity is likely to occur. Second, because any relationship between job insecurity and industrial relations is most likely to be manifest where job insecurity is a *collective* or widespread experience rather than predominantly an individual one. This is not to suggest that job insecurity *necessarily* occurs where decline exists but that it is a likely setting in which it might arise (Greenhalgh, 1983; Greenhalgh and Rosenblatt, 1984).

Organizational decline has become a topic of interest since the recession of the early 1980s (Whetten, 1980a; Greenhalgh, 1983; Cameron et al., 1988). Restructuring and rapid organizational change ensures its continuing topicality. Decline is discussed at some length by Greenhalgh and Sutton in Chapter 8. Here, we note that decline can be defined as the maladaptation of the organization to its environment or as a reduction in organizational effectiveness. Maladaptation, it is argued, can be caused through change in the external environment or through internal weaknesses which erode performance in a stable environment – or as some combination of the two. There is a danger of viewing organizations as belonging to one of two types: well adapted and maladapted. Perhaps the truth is that any organization can go into decline where there is a sudden and rapid change in the environment to which the organization cannot adapt sufficiently fast to avoid performance decrement. The external trigger for decline might come from recession, high interest rates, unfavourable exchange rates and so on. The successful organization of one day may be in serious difficulties the next. The danger of the concept of organizational decline is that it focuses attention on the organization rather more than on the *relationship* between the organization and its environment. Nevertheless, it has value in defining the organizational processes which may be occurring.

If organizational decline is partly about a decrement in performance (whether internally or externally caused), then this can be operationalized in terms of various indicators of performance such as market, market share, productivity and so forth (Nash, 1983). In addition, it is argued here that one important indicator of decline (of value in the study of job insecurity) is that of a decrease in the size of the organization, here defined in terms of the number of employees. Loss of employment through redundancy within the last three years can be taken to be one unambiguous indicator of decline when combined with other market-performance indicators (on its own it would be insufficient since it could include organizational growth through technology investment for example). The loss of employment through redundancy provides a clear signal to the workforce that there are or have been difficulties.

Organizational decline may take different trajectories (Argenti, 1976; Whetten, 1980a). Argenti documents three types of organizational failure, relating these to different types and ages of organization. While this overall typology is flawed, there is value in the idea that decline may occur suddenly and rapidly or may consist of more gradual and steady decline. In either case, internal weaknesses may be exposed by a change in the external environment. Decline may occur in mature and successful companies as much as ones struggling to survive.

The point being made is that decline can follow different paths and the consequences for workforce and management may consequently be different. Maladaptation through gradual failure may result in different effects on job insecurity from that where a major or sudden change in markets through restructuring or recession occurs. There is research to be done on the relationship between trajectories of decline and job insecurity but that is not the purpose of this study. Here, we take one situation of decline and explore its consequences both for job insecurity and for industrial relations. The case study explores job insecurity in the organizational context of fairly rapid and sudden change.

The context of decline provides a situation in which job insecurity can be investigated. We cannot assume that job insecurity exists because the material conditions for it are present. We must assess the role and value of job insecurity as an intervening variable between the objective conditions of the organization and the industrial-relations processes and outcomes. Jick and Murray (1982) and Hardy (1985) in their studies of decline and closure in public sector organizations in the USA and UK have argued that not only the objective reality but also the way changes are *perceived* by organizational members is crucial to organizational, group and individual responses to decline (this is a point we will elaborate later in considering causal attributions of decline). Also, the foregoing chapters in this volume have emphasized the subjective nature of job insecurity and the variation between individuals in their perceptions and evaluations of uncertainty. So we would expect considerable variation between workers in their feelings of job insecurity in a setting of decline.

The concept of organizational decline is also important for its *direct impact* on industrial relations. In particular, decline can create severe constraints for management in the resources and time-scale within which decisions and actions must take place (Cressey et al., 1985; Guest, 1982; Greenhalgh, in Chapter 9 of this volume; Legge, 1988; Strauss, 1984; Purcell and Sisson, 1983). Management may look for changes in their industrial-relations strategy in order to survive, imposing fast and drastic solutions to labour utilization problems (Purcell and Sisson, 1983). Batstone (1984) notes this to be a high-risk strategy which may be employed to deal with crisis. Guest (1982) suggests that the styles of management during the recession of the early 1980s varied according to the resources available in the organization and the timescale in which industrial-relations changes were required by management. Cressey et al. (1985) suggest from their case studies of 'just managing' that managements have a range of options in the way that they respond to crisis, but that crisis may subvert and narrow the options perceived to be available.

The concept of decline is also important in the discussion of intergroup power. This is a debate to which we will return under the heading of union–management relations and power (see below).

Job Insecurity Outcomes: Trade Union Involvement
The first group of individual-level variables which may be affected by job insecurity concern union involvement. The existing literature has little to say about whether job insecurity weakens or strengthens workers' trade union involvement. Price and Bain (1983), for example, suggest that rising levels of unemployment nationally may either *encourage* or *discourage* union joining. This is because workers may join the union as an insurance against job loss or may feel that the union is insufficiently powerful to negotiate effectively for them during recession. Union membership in Britain has been declining at an aggregate level throughout the 1980s (Kelly and Bailey, 1989; Kelly and Heery, 1989), though some of this decline has been attributed to national changes in the composition of industry and the labour force. There are a variety of reasons why workers decide not to join or to leave a union (van der Veen and Klandermans, 1989), including organizational factors as well as their own propensity. The impact of job insecurity on joining or leaving decisions is not clear and requires investigation (see also Chapter 5).

In circumstances where a closed shop formally or informally exists (as in the case study) there is less scope for the individual to choose whether to belong to the union or not. However, there are other dimensions of union involvement which are of interest. Union involvement can usefully be thought of as multi-dimensional (McShane, 1986; Klandermans, 1986; Nicholson et al., 1981) and following these studies, union involvement in this study has been divided into: level of union commitment; willingness to be a lay (workplace or branch) official; and willingness to take industrial action.

Some of these individual-level consequences of job insecurity (such as willingness to take industrial action) have been discussed in Chapter 5, so the arguments will not be repeated here, though data will be given in the case study on these issues. In brief, given the prevalent assumption that 'fear of unemployment' has weakened trade unions, it is reasonable to hypothesize that job insecurity will lower union commitment and reduce willingness to take part in industrial action (either strikes or lesser action).

It is also anticipated that job insecurity weakens interest or motivation to become a lay representative or shop steward. Internal motivation to become a shop steward is less common than being pushed into the role (Nicholson, 1976). However, we would expect that job insecurity would reduce interest in union office-holding even further. Job insecurity is believed to reduce the attractions of office-holding both because of increased visibility of the role to management and because of the increased frustrations under conditions of organizational decline. However, it is interesting to note that Nicholson et al. (1981) found in their study of white-collar local authority workers that union activists were more likely than members to be concerned about their own job insecurity.

Job Insecurity and Causal Attributions
Earlier chapters in this volume have shown how the same indicators of organizational performance may be perceived and interpreted in a variety of ways by employees. In addition, organizational cues about performance may themselves be ambiguous. Also, management may conceal or misinterpret some of the more damaging evidence on performance (especially where candid information could have a negative impact on external groups such as bankers, suppliers and customers). Thus, the potential exists for considerable ambiguity and uncertainty – and scope for extensive social and personal interpretation. Job insecurity in the context of organizational decline may be analogous to a massive projective test for management and workforce alike, with plenty of opportunities for the projection of fears, anxieties, defences and other feelings.

In considering subjective interpretations of organizations, the approach here uses the concept of causal attributions. Attributions have been explored already in Chapter 3. Briefly, these are the beliefs a person holds about the causes of events (see Kelley and Michela, 1980; Harvey and Weary, 1984; Antaki, 1981; Hewstone, 1988, 1989). People vary in their explanations of the same event. It is also possible to talk about attributions occurring at the collective level (Hewstone, 1988), where a group can subscribe to certain explanations of particular events. This level of attributional analysis has been less fully explored. Social representation theory is also related through the idea of shared social beliefs and knowledge (Moscovici and Hewstone, 1983) and has been applied to the situation of temporary unemployment (Depolo and Sarchielli, 1986).

The interest in attributions lies partly in the hope that understanding the causes people attribute to events will help explain the types of behaviour they engage in. At the organizational level, there is also empirical support for the relationship between action and belief. Jick and Murray (1982) argue that responses to crisis in the public sector depend on how the crisis is perceived to have arisen. Although they do not use the term attribution, they examine the beliefs organizational members hold about sources of crisis and argue that these beliefs shape responses. They conclude that responses vary crucially according to whether the causes of the organization's difficulties are seen to lie inside or outside the organization. In our study we use this dimension, labelling these organizational or environmental causes. An example of organizational causation is where management is felt to have handled financial affairs so badly that crisis ensued. An example of environmental causation is one over which the organization is perceived to have little or no control, such as recession or government policies.

The dimension of organizational/environmental attributions can be used to help explain individual and collective attitudes and behaviour within declining firms. Where workers as a group perceive the causes of organizational difficulties to be outside the organization (environmental) there may be a stronger basis for co-operative attitudes and relations. Sherif's (1966) functional theory of inter-group conflict suggests that where two parties can align their efforts towards a common and superordinate goal, such as in this case preventing organizational closure, then relations can change from hostile to co-operative over a period of repeated interactions.

The implications for organizationally located attributions are more complex. Organizational causes (that is, inside the organization) may reside primarily with one's own group or with the other party. The implications are different. For example, if workers as a group perceive organizational difficulties to be caused by their own actions (for example, high wages, low effort, restrictive practices), there may be some basis for co-operation to avoid further organizational problems. However, a belief that organizational difficulties are due to the other party (for example, incompetent management or poor organizational decisions) provides little basis for co-operation and indeed may predispose the parties to conflict. Of course, given that we are exploring a complex event (decline) which may have several causes, attributions may also be multiple – and hence *both* organizational and environmental. We can, however, assess the balance of causations, or the predominant influence on attributions.

Although we suggest that attributions may affect inter-group behaviour, there needs to be a recognition that group processes can affect attributions (Brown, 1988; Hewstone, 1989). Research in other contexts has shown a strong tendency for groups to attribute the cause of *negative* outcomes to the other party rather than to themselves (and the cause of *positive* outcomes to their own group). Given that organizational decline is a negative event one may expect a stronger emphasis on other-party rather than own-party causes where organizational attributions are employed. In other words, group processes may influence attributions as well as vice versa.

So the questions for the research to ask about attributions are twofold. First, does job insecurity affect the type of attribution made? That is, on an individual level, do more insecure workers use different explanations of organizational difficulties than their securer colleagues? Second, do workers as a group use a particular type of attribution (organizational/environmental) and does this relate to union–management relations in the organization?

Job Insecurity and Trust
Trust, following Cook and Wall (1980), is taken to consist of two factors: belief in the *abilities* of the other party and also faith in the *intentions* of the other. The measure here has been adapted to explore inter-group rather than interpersonal trust. We expect to find a close relationship between trust, job insecurity and inter-group relations. We suggest that insecure workers will have lower levels of trust in management (on both factors). In addition, at the group level, low trust will be associated with more other-party (organizational) attributions and with more antagonistic union–management relations. Cressey et al. (1985) suggest an organizational basis for this, although they do not use the concept of trust. They suggest that managerial legitimacy and authority is eroded by crisis: the proposition of managerial competence is not at face value supported by the existence of organizational difficulties. So one might expect levels of trust to be lower in an organization in decline than in one in growth. We may also note that low levels of trust are indicative of poor inter-group relations (Sherif, 1966; Fox, 1974). Kelly and Kelly (in press) suggest that trust levels may be reduced where management imposes change. Strauss (1984) suggests that the quality of union–management relations (conflict or co-operation) following cutbacks and recession will be affected by pre-existing levels of trust.

Job Insecurity and Industrial-Relations Climate
Finally, in this section on individual and group attitudes, we expect to find a relationship between job insecurity and a less favourable industrial-relations climate. Those who are more insecure are more likely to endorse 'them and us' attitudes and to express negative views about the handling of industrial relations in the organization.

Job Insecurity and Union Organization

We turn now to look at intra-group processes within the union organization. Does job insecurity change existing patterns of union structure and behaviour? We will examine the stability and structure of the workplace union organization, which consists primarily of the relationships between shop stewards and between stewards and members. We need to examine union organization before turning to the power bases and power relations existing between management and unions in the organization. However, the issues of intra- and inter-group processes are inevitably related, both in social psychological terms (Brown, 1988; Sherif, 1966) and because in industrial relations it has been suggested that union structure and organization is strongly shaped by collective bargaining and the actions of the employer (Clegg, 1976; Terry, 1983).

There has been considerable interest and debate as to whether recessionary conditions have affected and weakened union organization at plant level. Commentaries at the beginning of the 1980s suggested that trade unions had been or would be devastated by the recession and high levels of unemployment (Lane, 1982; Winchester, 1983; Bassett, 1986; Towers, 1982; Rose and Jones, 1986). The argument, as we have already seen, was that high levels of unemployment would reduce union members and bargaining strength. The intervening variable of job insecurity or fear of unemployment was largely assumed and never investigated.

There are theoretical reasons for rejecting such a deterministic view of the impact of unemployment and recession on union organization as we have seen. In addition, there is evidence, based on British industrial relations, which bears on the question of union organization in a period of change. This evidence is also important internationally as we shall explain. Several surveys of industrial relations at the workplace and in organizations have been conducted during the 1980s (for example, Edwards, 1985a, b; Marginson et al., 1988; Batstone, 1984; Daniel and Millward, 1983; Millward and Stevens, 1986). Summaries of these surveys are available in Terry (1986) and especially Legge (1988). Although the concept of job insecurity is not included as an intervening variable it is implicit in several of the surveys.

Overall, while the evidence is complex and varies by organizational size, industry, public or private sector and so on, it suggests that recession and restructuring have not led to simple or dramatic reductions of union organization, strength or scope of bargaining. Changes in national statistics on union recognition, membership and density levels, structures of organization and bargaining scope can be largely attributed to changes in the industrial base rather than to changes within organizations (Legge, 1988; Batstone, 1984; Marginson et al., 1988).

The survey evidence, especially that available across time (for instance, Millward and Stevens, 1986), shows that union *structures* have remained relatively intact within organizations despite the existence of widespread unemployment. At workplace level, union recognition and union density have not changed dramatically. The number of workplace union organizations having regular steward meetings has declined slightly but can partly be accounted for by the decline in the absolute numbers employed, resulting in less need for co-ordination (Legge, 1988). Overall, the evidence confirms a picture of little disturbance to institutional arrangements despite the existence of widespread redundancy. However, the public sector is a notable exception to this, where government policy has been hostile to union growth and union bargaining.

Of course, this evidence concerns the UK only. However, it can be argued that if union organization – in structural terms – has not been greatly disturbed here, with a hostile government, a strong economic downturn and union organization based particularly in the workplace, then this presents a strong case for the proposition that union organization has not declined due to the fear of unemployment. If this applies in the UK then there is a basis for believing that other countries, with fewer pressures on union organization, may also have kept union structures intact.

However, as Legge (1988) notes, structures of union organization may remain intact while failing to retain the vigour of earlier activity (the 'empty shell' hypothesis). Terry (1986) observes that knowledge of the formal structures is insufficient to reach conclusions about union activity and strength.

We suggest here that union organization can be additionally assessed through recourse to some consideration of the feelings of job insecurity among shop stewards and members. Where union members do not feel threatened by the possibility of job loss, then union organization may be unaffected but we expect that feelings of job insecurity may make employees more cautious about standing for the position of shop steward.

In addition, at the level of intra-group processes, we would expect that job insecurity would reduce group cohesion. Indeed, it may well increase sectionalism in the union organization. Sectionalism is here defined as the domination of workgroup or craft-group concerns over more general concerns for the union as a whole in the workplace. Sectionalism is typical in certain industries, notably engineering, whether union organization is weak or strong (Terry and Edwards, 1988; Brown, 1973; Batstone and Gourlay, 1986) and is often associated with a fragmented payment structure or strong craft identities. However, we suggest that the experience of job insecurity may increase sectionalism because declining resources, including jobs, in the organization may increase competitive pressures between different sections or departments. The assessment of the degree of competition or co-operation within the union organization is a most interesting one in the context of organizational decline and job insecurity.

Job Insecurity and Union–Management Relations

At the heart of discussions about the impact of restructuring has been the question of union–management relations. Has the existence of high levels of unemployment given management more power? Has recession created a new style of management which is brutal and assertive, the so-called 'macho management'? If the period of full employment of the 1960s and 1970s was a time of union power in the workplace (Donovan, 1968; Batstone and Gourlay, 1986) does that mean that management have been able to 'reassert control' over industrial relations in the leaner 1980s? What has happened, if anything, to the union power base? Does this set a pattern for the future?

We can set these questions within two frameworks which may help us to interpret outcomes and processes of union–management relations. The first is inter-group behaviour. The second involves a detailed consideration of power. However, first we will briefly review some of the evidence on union–management relations and restructuring. Again, reference to job insecurity is by implication rather than by direct expression.

Early in the recession, the idea that fear of unemployment caused a weakening of union power and a moderation of union demands was popularly asserted, with a scenario being drawn of much weakened or even irrelevant trade unions and with management having the stronger hand (for example, Ogden, 1981; Bright et al., 1983; Kelly, 1984; Hunter, 1980).

Whether management power was used coercively, through 'macho management', or whether managements were now in the comfortable position of simply being confident enough to bypass unions were both offered in these analyses. However, it was soon realized that notions of management coercion were over-generalized from a few prominent cases (Brown, 1986; Martin, 1987; Batstone, 1984; Legge, 1988). Instead, a more complex analysis was forged which was less deterministic and less focused on a simple shift of power to the employer. It was argued that recession also weakens the employers' position because of the highly competitive and declining product market. An employer may not wish to risk opposition or a dispute, with disruption to production, by the assertion of managerial authority. The employer, in fact, has an incentive to win co-operation from the workforce (Cressey et al., 1985). Alternatively, management's weakened power base could impel them to look for changes in their industrial-relations strategy simply to survive. Then fast and drastic solutions could be imposed on labour-utilization problems (Purcell and Sisson, 1983). Strauss (1984) notes that managements may no longer have the resources to pay their way harmoniously into changed working practices and relationships. We can note the basis here for different interpretations of union–management relations and restructuring.

The weakened state of both parties could in fact lead to increased cooperation in some cases (Purcell, 1981; Strauss, 1984; Brown 1986; Cressey et al., 1985), through both parties' recognition of extreme threat. Cressey et al. suggest that there is no simple managerial response to recession. Managements have a range of options in the way they respond to crisis, although crisis may subvert and narrow the options perceived to be available. The survey evidence cited earlier also notes little major change to union–management relationships in terms of formal structures: bargaining arrangements and bargaining scope show little evidence of change. However, both Legge (1988) and Terry (1986) sound a word of caution here, since this could represent the 'empty shell' phenomenon, or the bypassing by management of formal bargaining, through for example joint consultation or direct communications with the workforce.

Although the early work, then, suggested a weakening of unions relative to management, later work has suggested a range of options. However, little has been suggested in the way of intervening variables. Some understanding of the social psychology of inter-group behaviour could be very valuable here (see also Kelly and Kelly, in press). Where there is a perception of common interest or a superordinate goal, then co-operation could be facilitated (Kelly and Kelly, in press; Sherif, 1966). Increased co-operation would not be automatic, however, since inter-group attitudes, levels of trust and industrial-relations climate might prevent co-operation occurring to any significant degree (see also Strauss, 1984). Thus, subjective interpretation of conditions and the feasibility of co-operation may be more salient than objective conditions. The role of job insecurity and the attributions of organizational difficulties could be important in shaping union–management relations.

It is also possible that conflict could occur between the parties. Where attitudes are negative or hostile, where trust levels are low and attributions blame the other party, and where the organizational context causes a squeeze on resources, then there is little basis for expecting that union–management relations will be harmonious or conflict-free. Greenhalgh (1983) suggests that job insecurity in the context of organizational decline will lead to increased conflict between the parties as each experiences greater constraints than previously.

The current research, based on one case study, is not concerned with developing hypotheses about degrees of conflict or co-operation across different organizations according to the state of inter-party processes.

However, we do provide a framework for understanding what may be happening within one organization and this could be extended in further work.

Although inter-group processes are important in shaping union–management relations, there is also the need to consider power between the parties since this will be a further factor affecting levels of co-operation and conflict within organizations. Assessment of union–management relations in the context of decline and insecurity cannot be made without recourse to the concept of power.

Yet within the literature about industrial relations and recession, the conceptualization of power has been inadequate (see also Kelly, 1987). There have been attempts to assess the power of each party using indicators solely of the party itself. While that may help us understand the power *base* of the parties it does not tell us about power *relations*. For example, union membership and density levels have been used to argue for trade union power in a period of high unemployment. Terry (1986) extensively surveyed the evidence on indicators of union power such as industrial action, bargaining scope and job controls. However, these cannot be interpreted without also understanding managerial power bases. The failure of industrial relations to analyse power adequately may explain why researchers have been at such a loss to understand the effects of recession. For example, Terry (1986) raises the difficulty (but is unable to answer it) of how to assess power where the interests of the two parties converge rather than conflict. Kelly (1987) notes several examples where an assumption of conflicting goals is not only inadequate but misleading. For example, the argument that union strength can be read from the size of the annual wage increase is specious because management may have awarded a large increase for reasons unconnected with union power (such as to motivate or to reward flexibility). For this reason any conceptualization of power, we argue, must consider *goals* rather than simply *outcomes*.

Weber defines power as 'the probability that one actor within a social relationship will be in a position to carry out his will despite resistance' (1947: 152). Although this has been a popular definition for social scientists to use, we reject it here because of its emphasis on a conflict of interests between the parties. It implies that what one party loses the other gains: a zero-sum view of power. Instead, we prefer the view, advocated by Kelly (1987) and Kirkbride (1985), that an approach to power must allow for convergence as well as conflict. Here, we find Hyman's definition helpful, which is 'the ability of an individual or group to control his (their) physical and social environment' (1975: 26).

Such a framework takes a non-zero view of power. It is quite conceivable that both parties may be unable to achieve their objectives. Thus they may both be weakened by the situation of organizational decline: for example, the management by the state of the product market and the union members by the state of the labour market. This is because for each their *power base* has been weakened. Other outcomes are also possible. This approach to power is in complete contrast to the zero-sum view, which has been prevalent in discussions of industrial relations in recession.

The current approach therefore takes an exchange perspective on power (for example, Bacharach and Lawler, 1980; Pfeffer, 1981), where power is seen to be the property of a *relationship* rather than an attribute of a party. While each party has a power base, this exists to the extent that the other party is dependent on something which the first party can offer. And the power base may be weakened by the inability to mobilize their power resources. Also, in this consideration of power a sanction must be available or be believed to be available if one party is to influence the other. So in considering the power resources and the power relationship between unions and management in the context of organizational decline, we focus particularly on the goals, dependencies and sanctions of each party.

In relation to union goals, job insecurity might be expected to have several effects. One union goal may be to preserve jobs as far as possible. Yet given the intensely personal and highly subjective nature of job insecurity, it may be difficult for a union to develop a collective approach to stemming job insecurity. Indeed, job insecurity may well reinforce any existing pressures in the union towards sectionalism. Job insecurity may divide workers more than it unites them. As a consequence some of the potential power base of the union will be dissipated since it will be more difficult for the union to decide on what its goals and desired outcomes are. Batstone and Gourlay (1986) note that sectionalism weakens a union's strength to pursue particular outcomes.

Job insecurity is expected to have other effects on union goals. Where workers and unions believe that other goals can only be gained at the expense of secure employment, they will have to calculate the extent to which they wish to pursue those other goals.

Management, for its part, has its own goals, but also dependencies and sanctions. The literature already reviewed suggests that management, far from simply gaining in power as a result of recession, may in some cases be *more* dependent on the skill and goodwill of the workforce. Product market difficulties may place constraints on managerial action. Thus the union may be able to exploit that dependency in certain areas. Guest (1982) notes that in companies in recession financial considerations may outweigh personnel or employment objectives. This may result in some changes in the goals, dependencies and sanctions available to management.

It is suggested that there may be a set of different outcomes for particular organizations depending on the extent to which unions and management are able to pursue their objectives and goals. It is quite conceivable that each party may either gain in strength or lose its power resource as a result of recession and restructuring. Thus there are a set of possible outcomes in terms of union–management relations. While it is possible to examine outcomes such as industrial action, any full analysis of power relations will have to consider what each party is trying to achieve. While job insecurity is likely to weaken trade union power, it is also true that conditions of decline will weaken the management. Whether this situation results in increased co-operation or increased conflict will depend, it is argued, on the inter-group attitudes which exist and the attributions which are made about the causes of current difficulties.

Summary

This chapter has examined some of the existing macro-economic and institutional literature on industrial relations and organizational restructuring and suggested that job insecurity is a valuable intervening variable to explain industrial-relations attitudes and behaviour. In this chapter we put forward a framework which acknowledges the economic and organizational constraints within which industrial relations take place but also gives due attention to subjective interpretations of reality in industrial relations. Job insecurity has been identified as one variable which is implicit but unexplored in the literature on industrial relations. The impact of job insecurity is explored here both for its impact on individual attitudes and behaviours and for its influence on collective behaviour. A number of propositions are put forward. These will be explored in the following chapter where a case study of an organization experiencing rapid change and decline is described.

7

Industrial Relations and Job Insecurity: Learning from a Case Study

Jean Hartley

This chapter uses case-study data from an organization in decline to explore the propositions about decline, job insecurity and industrial relations outlined in the previous chapter. We begin by describing the company in decline and looking at how this affected organizational functioning. We then turn to a social psychological level of analysis, focusing on employee perceptions of job insecurity and how these varied. Then, the consequences of decline and insecurity are together examined for their impact on workplace industrial relations, including union organization, union–management perceptions and relations, and an assessment of the power of the two parties.

The data are drawn from a single organization, with a questionnaire sample of 137 predominantly manual workers (response rate 42 percent), over 100 interviews with fifty-one people across all functions and levels, and an intensive six-week period of observation, including attendance at union and union–management meetings. Documentary material from the company was also used. The data were collected in 1984 and 1985.

At first sight, it may appear that the case-study company is unusual in its serious crisis and its attempt to survive. However, the case study can be more usefully read as a more extreme example of everyday and familiar processes of change, uncertainty, insecurity and industrial relations. Its situation is not dissimilar from that of the Dutch companies described in the earlier chapters. Although the study took place during a period of recession, we have already argued (Chapter 1) that change, restructuring and environmental uncertainty are likely to continue for some time, so that some of the pressures on organizations and on individuals will be similar. Furthermore, although the crisis of survival creates additional pressures on management to act with short-term expediency (and this will be evident in the case study), in fact it has been widely argued that British industry in particular and Western organizations more generally have often been characterized by pragmatic management and industrial-relations firefighting (Terry, 1983; Purcell and Sisson, 1983). Thus, the case study may help us see more clearly and starkly issues of more general relevance to industrial relations and change.

The Company in Decline

The concept of decline has been discussed elsewhere (for example, Whetten, 1980a, b; Cameron et al., 1988; Argenti, 1976) and has been explored in the previous chapter. Various trajectories have been described, from sudden changes to more gradual, long-term loss of markets and effectiveness. Here, we describe a company which experienced rapid shifts in its fortunes after a long period of success. The company manufactured industrial vehicles. It had three sites, all in the same town, which were responsible for the machining, assembly, and parts and service respectively of the production process. Head office was located adjacent to the assembly site. The company produced expensive but well-designed products of high quality. A range of models, with a large array of modifications, were available to meet customer specifications. Assembly was one-off and small-batch production – essentially this was 'bespoke' engineering. At the machining plant, components were made for assembly and spares. There was a high level of production integration, especially between assembly and machining.

In 1984, there were 1250 employees, of whom just under 500 were in manual grades. The manual workforce was predominantly male, skilled and completely unionized, with membership being divided fairly equally between two unions. There were twenty manual shop stewards who cross-represented the shop-floor. The company belonged to the Engineering Employers' Federation and therefore had established procedures for collective bargaining and dispute resolution.

Decline became an issue for the company with the onset of recession in 1979. However, for a long time in the 1960s and 1970s the company had thrived, being seen as the British leader in its sector. Expansion and growth occurred through takeover and acquisition. First, it had been bought by a large multi-plant enterprise. Then, just before the onset of recession, the company itself acquired two firms in its own business area, which led to a doubling of size and an expansion of its markets. In 1980, it employed around 2000 people.

The climate of expansion was abruptly halted by recession. The organization experienced difficulties in adapting to its environment due to changes in the *external* environment. Recession meant a severe drop in the overall UK and world market of 40 percent between 1979 and 1981. In addition, performance deteriorated due to factors *internal* to the organization. It was losing its share of the already shrinking world market to Japanese and European competition: market share had fallen 6 percent in the two years prior to recession and continued to decline in the 1980s. A senior management report in 1980 pinpointed a number of internal problems with performance: high production costs, poor availability of vehicles, an expensive product, let down by poor finish;

a high level of rectification and indifferent after-sales care; a failure to develop basic data for plant loading; high inventory; a haphazard components-ordering system; an over-optimistic sales force; and the lack of an effective or strategic personnel and industrial-relations policy.

The combination of internal weaknesses and external changes had a dramatic effect on the fortunes of the company. Performance indicators slid downwards. The shrinkage in market and market share meant a halving of production. Gross profits dropped from 13 percent in 1975 to 2 percent in 1979. The order book lead time fell from over a year to around two months. Performance on all these criteria remained low throughout the 1980s. The organization was clearly struggling to survive.

In 1986, a year after the research, the company had such severe financial problems that it was put into receivership.

In employment terms, the period from 1980 onwards was characterized by high levels of uncertainty as a result of several waves of redundancies and a change in ownership. Reorganization of production spelled major job losses. One acquisition was sold. The parent company management was undecided for several months whether the case-study site or the second acquisition should be closed. This generated considerable anxiety and job insecurity among the workforce. Finally, the acquisition was shut down. This was not the end of the troubles. A decision to outsource much of the work at the machining plant led to around 150 'voluntary' redundancies, or two-thirds of that workforce. In early 1983 there was a further call for 180 volunteers, mainly in the indirect and staff areas. Later in the year came further redundancies. Even in 1984 the search for cost-cutting continued, with small pockets of work being modified or reduced. Throughout 1984 the redundancy scheme was open to anyone who wished to discuss terms with the personnel department.

A change in ownership also generated concern among all employees. In 1981 the parent company sold the firm to a consortium of shareholders, the major of whom had a controlling interest in the principal British competitor. Speculation was rife at all levels in the company and in the trade press that this heralded a merger in the near future. In the event, this did not happen although it was a chronic source of concern to workforce and management throughout the five years to 1986 that ownership continued.

Overall, we see a medium-sized organization experiencing considerable difficulties, as it contended with a worsening external environment and the internal weaknesses highlighted by these circumstances. The workforce had experienced several waves of redundancy and a change in ownership.

Organizational Functioning

A description of the functioning of this company undergoing rapid change is essential for understanding many aspects of job insecurity and industrial relations. It also provides a view of an organization in crisis, which is not common in the organization-studies literature. Even where retrenchment has been described this is often of large organizations with reasonable financial and other resources where change has been relatively ordered and well planned over a period of time.

Financial problems, with major losses from 1979 onwards, were exacerbated by high interest rates and the removal of the financial umbrella provided by the parent company in 1981. Under consortium ownership, financing debt became highly problematic and was a source of regular discussion between senior management and the company's bankers. Cash flow was a major preoccupation and sales revenue was used immediately to purchase supplies. As an increasing number of suppliers refused credit, some supplies could only be obtained when paid for. This intensified cash-flow difficulties and affected production.

Stop–go production, or 'storming', resulted from the intermittent and unpredictable parts supply. In the early part of each month, there was only a moderate level of activity because of the shortage of purchased components. However, the later stages of the month could be very pressured and chaotic. The assembly programme was modified as components failed to arrive and as the finance department urged the bringing forward of certain orders for reasons of solvency. Towards the end of the month extra effort was urged to push as many vehicles through to despatch, in order to create a favourable sales ledger and impress the bankers. The sales department compounded difficulties by promising short delivery dates to customers to gain sales. The irony for the workforce was that despite the earlier redundancies and current periods of inactivity the company regularly required high levels of overtime, which ran at 12 percent of total hours.

Additionally, the parts shortage resulted in efficiencies in the use of labour because workers were asked to improvise to deal with crises or shortfalls. This took a variety of forms but most galling to the workforce was the 'cannibalising' of finished or part-finished vehicles to obtain parts for another order which was deemed more urgent. Rebuilding and more rectification work then followed from this. Overall, stop–go production meant that headcount levels were higher than with smoother production. Management reported that value added per direct worker was less than half that for the industry as a whole.

During the redundancies of the early 1980s, many managers had left. While some new managers had replaced them the overall effect was a team with short job tenure. Key positions were held by managers without knowledge of the industry. Fragmentation characterized managerial action. Lateral integration was poor. Each management function became preoccupied with its own objectives and targets and pursued these at the expense of corporate objectives. For example, the sales function offered short delivery dates to customers to gain sales, but these interfered with production planning. There were wrangles between finance and production over the build programme for the month – the implications for the sales ledger, for production efficiency and for the workforce bonus often diverged. There were numerous occasions when production objectives were pursued at the expense of industrial-relations considerations, as we shall see.

Fragmentation was also increased by immediate crises. Day-to-day firefighting drove out longer-term planning. Where plans were made, they often disintegrated. For example, each month the appropriate managers met to plan targets and priorities in production. With parts shortages, customers changing specifications and the finance department urging the bringing forward of certain orders to stay solvent, plans soon degenerated. The reality was a daily production meeting between finance, production and purchasing, held in an atmosphere of controlled panic, to deal with the next day's activities. Such meetings got more frantic as the month progressed.

Planning was severely inhibited, also, by the poor or non-existent management information systems. Productivity was difficult to calculate, costs and profit per vehicle were only recently available (until 1984 sales staff were selling the full range of vehicles without knowing which were the profitable models). Purchasing, recently computerized, was still coming to terms with the new system.

Despite the serious organizational problems, the chief executive attempted to impose some order and financial stability. Legally, the company had been separated into three (based on the plants) in order to impose some financial discipline over costs (though this had also increased problems of lateral integration). The personnel department had been urged to reform the complex payment system.

In 1985, a new vehicle was launched with a simpler design and reduced material and labour costs to be more price competitive. Development resources had been minimal and so the venture was a calculated risk. It was to have effects on the perceptions of job insecurity and on industrial relations.

The Experience of Job Insecurity

In this situation of decline, how did the workforce feel about job insecurity? By job insecurity, we mean concern over the future existence of one's job (see Chapter 2). Some of the quantitative data about job insecurity from the British study have been analysed in Chapters 4 and 5. Here, we will simply recap that the response to the question

'How secure do you feel in your present employment?' showed a considerable spread of beliefs. Five percent felt very secure and 32 percent felt fairly secure. Forty-four percent felt somewhat uneasy while 10 percent felt fairly insecure and 9 percent felt very insecure. So around a fifth of the manual workforce report feelings of insecurity and nearly half expressed some unease. In view of the organizational crisis and the liquidation of the company just over a year later it is perhaps surprising that so many workers report feeling fairly or very secure. We explore some reasons for this below.

The variation in feelings about insecurity is reinforced in interviews. Some manual workers ($n = 35$) described their perceptions of and feelings of security or insecurity. We grouped their responses into four categories, which ranged from very insecure to confidently secure. In effect, these are a continuum, but responses have been grouped into four categories in order to identify the variety of views.

Some workers expressed *considerable insecurity*. For example: 'All the time we have the prospect of redundancies looming over us,' and 'Things have been going downhill steadily in this company. People feel they have to keep their heads down in order to keep their job.' The sense of insecurity was chronic, having developed over a period of years as the company had waves of redundancies. Some felt that, whether the company did well or badly now, the result would be the same: job losses. Either the company would continue to decline or financial improvements would enable more efficient use of labour. The threat of merger was also ever present given the company's ownership.

The largest group of workers were those expressing *confusion and uncertainty*. They did not have an immediate sense of insecurity but they were unclear about the future of the organization or unclear about how to interpret organizational data, including information from management and the widely circulating rumours: 'We are uncertain about the way the company is going. There is always a parts shortage. It makes you wonder where the company is going and whether it will survive,' and 'Takeover is a new fear and there are lots of rumours about this. But when you're on the shop-floor there are so many rumours and no one is sure which ones are right and which are wrong,' and 'The future of this place is in the lap of the gods. At the moment we are holding our own but you just can't tell.' Behind such comments lay a constant watchfulness. Every change was weighed up for its possible outcomes for jobs and working arrangements. Rumours were analysed for their implications for jobs. Even apparently positive events were scrutinized pessimistically. For example, the launch of the new vehicle might result in less work in certain areas and therefore job losses. Workers reported that their perceptions of job insecurity changed from week to week as rumours and information changed. Job insecurity here is not a clearcut phenomenon but a shifting evaluation based on unclear or ambiguous data.

Other workers expressed *resignation*. For example: 'This place is always meant to be closing down and it hasn't happened so far. So now I just take it as it comes,' and 'Everyone here is depressed, but then that's the state of the country generally. If you don't work here then where would you go? Anywhere else is just as bad.' Such comments suggest a fatalism about the changing fortunes of the company. These workers do not indicate positive feelings of security. There is little confidence in the future but also an awareness that things could be worse.

Finally, some workers expressed *relative security*. For example: 'The company as you know is on the up. I haven't been concerned about the security of my job for a couple of years now,' and 'In this section we are fairly confident at the moment. We can't be trimmed down any more. We're at the bare limit of the workforce so no one is worried.' These confident workers are familiar with feelings of job insecurity but it is seen to be a feature of the past. Current views are strongly shaped by remembrance of the recent past, where 'People were demoralized. Closure was imminent. Everyone was in a terrible state. It was depression eight hours a day here. It was not an enjoyable period at all. I got to the point where I actually hoped the gates would be shut when I got here.' So it is perhaps less surprising now that a sizeable group of workers see their current circumstances favourably. Past experience may be an important ingredient of current feelings of insecurity.

Sources of Concern and Confidence

The personal and occupational bases of variation in these feelings of job insecurity have been described in Chapter 4 (Table 4.3) and can be briefly recapped. Age, seniority and number of dependants and wage-earners were moderately correlated with insecurity (Pearson correlations of .18, .12, −.17 and .16 respectively). However, interestingly, section and plant were not related ($\chi^2 = 2.15$ and 2.67 respectively, not significant statistically).

The *same information* was used by workers to support either their pessimistic or their optimistic assessments about jobs in the future. For example, the new vehicle was a source of confidence for some: 'If a bigger volume of the new vehicle is built then there will be plenty of work for us in assembly.' While for others it spelled work reorganization and job reductions: 'There is a problem coming up with the new vehicle because I understand it will need less machining and so that reduces our workload. So there are doubts and fears building up over that.' Financial data about the company were also a source of both confidence and concern. Some viewed the rumours of losses as extremely grave: 'This year the financial situation is not so bright again and so we have some insecurity again. I don't know why exactly.' While others were more optimistic: 'This company is improving financially and soon we will be completely out of the woods.' Others were frankly sceptical of the figures: 'Management say we made a £4½ million loss this year but that must be the biggest lie ever.' Similarly, information about manning levels, about overtime, about the parts shortages, about the sales orders, about possible new contracts, could be interpreted in diametrically opposite ways. Thus again, we see that job insecurity is not a straightforward reading of objective organizational conditions but is a subjective experience reinforced or shaped by information which can be interpreted in varied ways.

Interpretation was not helped by the failure of management to provide regular or systematic information either to the workforce or to the trade unions. So inevitably there was a high level of rumour. We have also noted that there were workers for whom the information was unclear. Often these workers did not know what to think or changed their views. Again, some initiatives by management to provide information might have reduced uncertainties and rumour.

Management's Perceptions of Workers' Job Insecurity

Managers were aware that job insecurity existed among the workforce. For example: 'There is no question at all about it. There is tremendous uncertainty currently for the workforce' (managing director), and 'They are always suspicious about job losses and they are not entirely certain they have finished' (production manager). They also saw that job insecurity generated concern about working arrangements and any changes being introduced: 'The company tells of its plans for development but the workforce thinks of the changes in terms of job loss' (sales director).

However, this awareness of job insecurity was not translated into action. For example, management planned changes for the new vehicle build but did not communicate their plans to the unions or the workforce. The changes were not the focus of formal union–management discussion or negotiation prior to implementation. Management expressed frustration that the workers seemed to fail to see the importance of change but focused instead on their jobs.

So there was a major discrepancy between management's drive for change and worker concern about possible job loss. As Greenhalgh and Sutton argue in Chapter 8, a failure of management to see the *significance* of worker job insecurity can exacerbate existing problems in the organization, especially preparedness or resistance to change. The failure to perceive job insecurity as a crucial element of change also contributed to levels of trust, the industrial-relations climate and the overall quality of industrial relations in the company.

Job Insecurity, Union Involvement and Union Organization

The quantitative data on the relationship between job insecurity and measures of union involvement and commitment were analysed in Chapter 5. There was no statistically significant relationship between insecurity and holding a union position (now or in the past, though the sample of office-holders was small; $\chi^2 = .90$). Nor was insecurity related to level of union commitment (Pearson correlation, $r = -.02$); which union the worker belonged to ($\chi^2 = 3.90$); or level of union involvement (from reluctant member to activist; Pearson correlation, $r = .02$); or interest in the union ($\chi^2 = .81$). However, more insecure employees were more likely to believe that union strength in the workplace was lower ($r = -.23$). See Table 7.2, page 135 for a summary.

The interviews also failed to find a relationship between insecurity and union involvement. Shop stewards' views of job insecurity varied as much as the rank and file. However, the shop stewards were more concerned than their members with the *implications* of job insecurity. They were particularly attentive and watchful for information and signs about company performance and job insecurity. It was the stewards who were most aware of the 'Catch 22' situation whereby either improvement or deterioration in company performance might presage job losses. So vigilance was high among them all.

Organizational decline and job insecurity had together played havoc with union organization in the company. During the redundancies, there had been a high level of shop-steward turnover. The general view in the organization, supported by the long-serving personnel manager, was that a disproportionate number of stewards had been made redundant as certain managers used the cutbacks to 'weed out trouble-makers'. As a consequence, around half of the twenty shop stewards had been in post for a year or less, and 'We have a floating population of shop stewards because many don't keep the job [of steward] for long.' The personnel manager commented: 'The whole shop-steward network has fallen apart in the last three to four years. Stewards kept being put on the redundancy lists and so now no one will take the job on because they know they are vulnerable. This means the choice of shop steward is even more limited.'

By contrast, prior to the redundancies, union organization had been highly stable. This was especially noticeable for the union leadership in the company. Before the job losses, the two union senior shop stewards (one for each union, working together) had been in post for over eight years each. As the redundancies got underway, there was a shift towards short occupancy of the role or even no one in post. The last four years had seen a total of seven senior shop stewards (including one removed from office through a membership vote of no confidence). Sudden resignations had occurred twice in the last two years. One current senior steward said: 'When the last senior steward resigned, I was made senior steward within two hours. I'd had no idea he was going to resign. He let me down badly. He just handed across his briefcase with documents in it. I had to find out what was happening when I arrived for meetings with management.'

The questionnaire confirmed a reduced willingness to serve as a shop steward. Personal motivation to become a steward is often low (Nicholson, 1976) so it is not surprising that 63 percent of the workers indicated no change in their interest in the role. However, 32 percent said that they were now less willing since the redundancies while 5 percent said they were now more willing. Willingness was not related to personal level of job insecurity ($\chi^2 = 1.12$, not significant; see Table 7.2 on page 135).

The recent lack of stability among stewards and especially senior stewards had serious consequences for industrial relations. Like the managers, the period of change and turbulence meant that union officers were in post with relatively little experience and low role continuity.

Lack of continuity created a number of problems for new incumbents, notably making information harder to obtain and interpret. It also made pre-existing sectionalism harder to overcome as strong central leadership was lacking. It was hard for the senior stewards to get an overview of the union organization. For example, the senior steward newly in post for three weeks during the research was unable to say how many shop stewards there were in the company or where they were located.

The interviews with stewards indicated the difficulties and frustrations of the role and the hesitancy in acting as a steward. We have already referred to the difficulties of lack of continuity. In addition, the existence of job insecurity also influenced role behaviour. Fear of victimization was clearly articulated. However, job insecurity was also manifest in more subtle ways. There were concerns about *job performance* (as opposed to steward role performance). One of the senior stewards explained: 'I will look after my own job because if I don't look after me I won't serve the union very well. It's important to me to stay in touch with my job – I don't want the company to say I'm never there and that my job is unnecessary. I don't want to risk losing my job. I need to be part of the factory as well as part of the union.' The other senior steward took a similar view: 'I don't like sitting in the union office all day . . . I want to be on my section a lot because if I'm away from the job too long management will say they don't need me.' In addition, informal pressures within a workgroup could influence the time spent on union work: 'There's so much work in this section that it's not fair on the others if I'm not here to give a hand.' Thus job insecurity operated through concerns held by individuals and groups to continue production as well as through uncertainties about how management might behave towards stewards.

The role of steward was also less attractive because of the increased pressures on stewards due to organizational decline and the day-to-day functioning of the company. For example, the expedient approach of management to introducing change and handling industrial relations caused a number of grievances and dissatisfactions among the workforce. A further difficulty was the poor union facilities. The union office was in a semi-derelict building a long way from working areas even for the plant where it was located. The telephone only worked intermittently. It was hardly a vantage point for union work, although prior to the redundancies it had been in the hub of activity by the canteen (now closed). The union office can be seen as a symbol for union activity at the company: left on the tide line as jobs ebbed through decline.

Sectionalism had existed well before the redundancies according to the full-time officials and the longer-serving managers. It is typical of much of the engineering industry. However, district union officials and company managers believed that sectionalism had increased with the pressures created by redundancy, job insecurity and the lack of continuity of senior stewards. Some stewards did not hold the senior stewards in high regard and most preferred to handle issues themselves rather than involve senior stewards. The senior stewards complained of secrecy and even deviousness on the part of the stewards; the stewards complained of interference and lack of impartiality from the senior stewards. Steward attitudes were somewhat contradictory. While not wanting interference, they were also critical: 'You don't walk around the shop and talk with us, you simply come in and tell us we oughtn't to be doing something or other.' So the senior stewards' role was made more difficult by their own hesitancy about the task and by the lack of support from other stewards.

Causal Attributions

We suggested in Chapter 6 that the causal attributions workers use in interpreting organizational decline may influence industrial relations, especially union–management relations. By causal attribution we mean the explanations an individual uses to describe the cause or causes of a particular social event. Here we examine the attributions given about company performance.

In the questionnaire, respondents were asked to rate each of eleven

Table 7.1 *Causal attributions of previous redundancies*
'Here are some reasons people have used to explain why redundancies happen. In your view how important was each reason in causing redundancies *in this plant* over the last three years?' (Four-point scale, very important to not important.)

Cause	% Rating very important	r^1	Attribution[2]
Poor decisions by managers at this plant	69	$-.18^3$	O
Lack of demand for products	64	.03	E
World recession	58	.07	E
Policy decisions by senior managers	55	$-.10$	O
Government policies	54	$-.10$	E
Management failure to get customers	48	.00	O
New technology	44	$.17^3$	E
Better performance by competitors	40	$-.02$	E
Not enough effort by workers	20	.10	O
Not enough co-operation from the unions	11	.08	O
High wages	8	$.18^3$	O

[1] Pearson correlation with level of job insecurity.
[2] Organizational or environmental attribution (classified by researcher).
[3] Statistically significant at $p < .05$.

possible causes of the previous redundancies in the organization from very important to not at all important (four-point scale). The range of causes can be categorized as organizational or environmental (see Chapter 6). Organizational causes are those due to events and processes inside the company such as poor management or low worker effort. Environmental causes are located outside the company such as recession or management outside the immediate company. Organizational attributions can be sub-divided into causation by own or other party. The results are given in Table 7.1.

An organizational cause, poor decisions by plant management, was the most frequently endorsed cause of job losses in the recent past, with around two-thirds seeing this as very important. However, an environmental cause comes a close second: lack of demand for the company's products. Overall, the table shows a mixture of organizational and environmental causes interwoven as the most frequent explanations of recent difficulties. Neither type of attribution predominates. It is also evident that workers do not see their own behaviour as contributing much to the redundancies. Only 20 percent believe that lack of worker effort was a factor and even fewer endorse other own-party causes.

We cannot comment on the accuracy of the attributions. Whether reflecting reality or not, they may form the basis for other beliefs and inter-group behaviours (Hewstone, 1989). Circumstantially, the Dutch

Table 7.2 *Relationship of job-insecurity measure with industrial-relations attitudes*

Variable	Pearson correlation	Chi-square
1 Union involvement (reluctant member to activist)	.02	
2 Union commitment	−.02	
3 Perceived strength of union	−.23[2]	
4 Attitude to handling of previous redundancies	.35[2]	
5 Trust in management	−.33[2]	
6 Faith-in-intentions subscale of trust	−.32[2]	
7 Confidence-in-ability subscale of trust	−.27[2]	
8 Organizational commitment	−.42[2]	
9 Job satisfaction	−.43[2]	
10 Conflict in last three years	.10	
11 Co-operation in last three years	.13	
12 Union–management relations in last three years	−.33[2]	
13 Industrial-relations climate	−.24[2]	
Impact of previous redundancies on		
14 own interest in the union		.81
15 tougher management attitudes to workers		6.14[1]
16 union strength		18.49[3]
17 union–management relations		15.98[3]
18 willingness to be steward		1.12
19 willingness to go on strike		7.39[1]
20 willingness to use lesser sanctions		2.40

[1] Statistically significant at $p < .05$.
[2] Statistically significant at $p < .01$.
[3] Statistically significant at $p < .001$.
χ^2 based on division of insecurity responses into 2 categories of high and low.

data (Chapter 5) suggested that attributions varied considerably between the companies studied and largely reflected the economic reality of the companies. In the British study here, we described organizational decline as arising from both external and internal pressures. The workers' attributions also reflect this complexity of causes. However, the absence of attributions involving one's own group but involving the other, in the case of a negative social event (such as decline), is also typical of parties in conflict or with high levels of mistrust (Hewstone, 1988).

In spite of our predictions (Chapter 6), few differences were found in attributions according to degree of job insecurity. Statistically significant, but moderate, correlations were found with three attributions (see Table 7.1). Insecure workers were more likely to believe that poor decisions by plant management were important in previous difficulties ($p = -.18$). This was the most important cause overall among the workforce. However, insecure workers were less likely to believe that new technology or high wages had contributed to past redundancies ($p = .17, p = .18$). Overall these results do not point to a distinction between secure and insecure workers and suggest that attributions are not as key to understanding insecurity as was anticipated in Chapter 6. The data cannot be compared directly with the Dutch study, where only insecure workers were asked for their attributions of their concern about their job future.

Although there is only a small difference in attributional style of insecure and secure workers, it is interesting to note that insecure workers were less satisfied with the way the previous redundancies had been handled. A four-item measure asked workers about their views on the process and impact of the redundancies ($r = .35$; see Table 7.2). While the questionnaire asked about previous difficulties, the interviews focused on attributions of current difficulties for the company. As with the questionnaire, attributions were not simple but were based on a set of causes. Environmental factors were widely cited, with Japanese and European competition and the world recession seen as important. Other environmental factors (such as government policy or changes in customer demand) were also mentioned by some workers.

However, organizational factors were seen as important too. It was widely believed that management were making a bad situation worse by their inability to run the organization effectively. Numerous examples and complaints were given of poor management and poor decision-making, which reinforced workers' views that the management was contributing substantially to present difficulties. Overall, then, from both questionnaire and interview data, attributions were broadly consensual, pointing to environmental factors exacerbated by organizational factors located in managerial incompetencies.

Inter-group Trust and Industrial-Relations Climate

Issues of trust were explored in both questionnaire and interview. The scale of Cook and Wall (1980) on interpersonal trust at work was adapted to explore inter-group trust, where the concept may be equally useful. The six-item scale used had two dimensions: faith in the intentions of management and confidence in the ability of management.

Levels of trust expressed by the workforce were low overall. The mean score was 13.91 with subscale scores of 6.91 for faith in intentions and 7.02 for confidence in ability. These are lower than those reported by Cook and Wall (1980) and Clegg and Wall (1981) after adjustment of means to take account of the five-point scale used here. The scores are virtually identical with the validating sample used by

Clegg and Wall in a white-collar service organization 'where levels of trust and trust scores were particularly low'. Illustratively, 65 percent felt that management would be quite prepared to gain advantage by deceiving the unions while 19 percent did not (intentions subscale). Also, 65 percent agreed that the company would have a poor future unless it could attract better managers (ability subscale).

Job insecurity was significantly related to trust. Table 7.2 shows that insecure workers were more likely to believe that management lacked ability (Pearson correlation, $r = -.27$) and they had less faith in their intentions ($r = -.32$). On the overall measure of trust, insecure workers trusted management less ($r = -.33$). This was also found in the Dutch samples (see Chapter 5).

Low trust was very evident in interviews. The management were perceived to be incompetent and manipulative (equivalent to ability and intention). Competence was judged from observations of the production process and the handling of industrial relations. A perceived inability to schedule work efficiently, to organize supplies, to prevent the cannibalization of vehicles, to plan ahead and anticipate problems were all strongly criticized by the workforce, who disliked what they saw as a disorganized, firefighting approach to production: 'Management have got everyone baffled here. There's a vehicle worth £20,000 which won't work and can't be finished because there is £50 worth of shortages. Yet they keep going on about how it is worth £20,000.' Trust in the ability of management to handle financial calculations on revenue, labour costs and bonus calculation was low. One steward commented: 'Each week I spend a couple of hours at least recalculating the bonus figures for the lads in my section. Then we have a couple of hours haggling over them [with management]. I have to do it every week because so often they get it wrong.' Overall, the view was that 'The shop-floor and the stewards don't trust the management here at the moment. There's a lot of doubt now, even by the people who normally accept everything. Everyone is more questioning. The management here is not competent to run the company.'

Manipulation and deviousness were also perceived by the workers. Not only was accurate information about finance hard to come by but to make matters worse they felt that financial and production information was adjusted to suit the circumstances. For example, workers believed that production figures were manipulated in the run-up to the pay claim with a deliberate inflation of the bonus.

One incident did much to damage management's credibility about finance. The finance department had produced a set of costings for the new vehicle axle assembly showing that moving the work to another department would result in a cost saving of £10 an hour. A stoppage of work by the aggrieved workers who were losing the work took place on the weekend prior to the well-publicized Monday launch of the new vehicle as the demonstration vehicles were still being completed. Vulnerable to such action, management promised to re-examine the figures. In a meeting of the senior stewards and personnel in the following week, new figures showed that the work was 40 pence an hour cheaper if it was built in the original section. While the stewards were pleased with this victory, they were well aware of its implications for other disputes: 'When the trade unions question the financial set-up, then the company plucks new figures out of the air.'

Deviousness was also seen by the workforce in the use of industrial-relations procedures. It was widely felt that the management had 'an amateur approach to industrial relations'. Stewards throughout the company believed that management were quite prepared to evade procedure when they felt like it: 'Industrial relations is a mess here. The management don't stick to agreements. They are always breaking them' (senior steward). Given the ambiguous and unclear nature of information in the company and its role in fuelling speculation about job insecurity, the low level of trust is a further contributor to deteriorating industrial relations.

Issues of trust were not explored in as much detail with managers but it was clear that trust in the shop stewards was also low. There was little regard for union representatives and little role envisaged for the unions in the fight for survival and in the adaptation necessary for the future. The shop stewards were seen as inexperienced. Criticism was particularly reserved for the two senior stewards: 'The two senior stewards are rather unsuitable. They've got no go and they can't control the membership. They don't understand documents and they both lack experience and competence' (personnel manager). Another manager saw the shop stewards as reasonable but limited: 'I've got fair confidence in the stewards, except on complicated issues. They tend not to understand certain issues and will pick out what they think are the important features and tell the shop-floor' (production manager).

All the managers recognized the limitations of their domestic union organization, especially the leadership. For some managers, the poor quality was due to a lack of confidence, experience or basic ability. Others saw it as a tough job in very difficult circumstances which required unusual skills and understanding. The stewards were criticized for having a short-term approach to problems, for failing to understand the gravity of the financial position, and for lacking the ability to deal with complex issues. These criticisms are almost exactly a mirror of the shop-steward perceptions of the management: being expedient, short-term and failing to consider the wider significance of events.

The questionnaire data add to the picture of distrust and a poor industrial-relations climate (with a four-item measure). For example, 87 percent felt that the company had a poor future unless management and workers co-operated. Three-quarters (74 percent) felt there was a widespread feeling of 'them and us' in the plant (only 11 percent disagreed). Over half (59 percent) felt that the workers had many long-standing grievances (compared with 16 percent who disagreed). Interestingly, there was a greater spread of attitudes where workers were asked for their opinion about whether there would be more co-operation in a more favourable economic climate: a quarter (28 percent) felt there would be more co-operation, while 36 percent felt there would be less. A further question asked if union–management relations had improved or got worse since the redundancies: around half (49 percent) felt that industrial relations had deteriorated, while only 5 percent felt it had got better.

These attitudes were related to the individual's degree of job insecurity. More insecure workers saw the industrial-relations climate as somewhat less co-operative or favourable ($r = -.24$ for the four-item scale, which is a moderate relationship and significant statistically). On the question whether union–management relations overall had improved or got worse because of the previous redundancies there was a significant relationship between insecurity and believing that relations had got worse ($\chi^2 = 15.98$, significant at $p < .001$). Insecure workers felt that union–management relations had deteriorated generally over the last three years ($r = -.33$). Insecure workers were also more likely to believe that the previous redundancies had made management tougher in their attitudes towards the workers ($\chi^2 = 6.14$, $p < .05$) and that the redundancies had made the union weaker ($\chi^2 = 18.49$, $p < .001$). See Table 7.2, page 135. We may note here too, in passing, that insecurity was related to lower organizational commitment (Pearson correlation, $r = -.42$) and lower job satisfaction ($r = -.43$) as shown in Table 7.2. It appears that there is a cluster of negative job, organizational and industrial relations attitudes which are associated with job insecurity.

Indicators of Conflict

So far we have seen that inter-group attitudes were broadly negative and that the industrial-relations climate was poor. What were actual relations like between the two parties? In the questionnaire, two questions asked about the individual's perception of the workforce's willingness to take industrial action since the redundancies. On strikes, 27 percent believed people were now less willing to go on strike and 14 percent thought people were more willing, while 60 percent indicated no change. So there was less willingness to take such action than previously. On lesser industrial action, such as imposing an overtime ban, there was a slight change in the other direction: 26 percent said that workers were now more willing, while 19 percent felt that workers were less willing and 55 percent reported no change. Although a small difference, it does suggest that it was easier for the workforce to contemplate action short of striking. This concurs with the broader survey evidence of the recession summarized by Legge (1988), which points to a greater actual involvement in lesser industrial action. However, the impact of job insecurity was not clearcut. Insecure workers were more likely to believe that workers in the organization were more willing to go on strike since the redundancies ($\chi^2 = 7.39$, $p < .05$) but were not more likely to believe that workers would take lesser industrial action (such as an overtime ban) ($\chi^2 = 2.40$, not significant). Also insecure workers did not believe any more than secure workers that the level of conflict and the level of co-operation had changed in the last three years ($r = .13$ and $.10$ respectively). See Table 7.2, page 135.

Industrial-relations outcomes, such as industrial action, are notoriously difficult to assess. Outcomes need to be evaluated within a broader consideration of the objectives and power resources of both parties. Here, though, we examine first some indicators of conflict through strikes, other action and external works conferences. These are interpreted more broadly in the following section where the objectives and power resources of the unions and the management are considered in detail.

It was difficult to obtain figures on industrial action for two reasons. First, redundancies in the personnel department had led to the abandonment of much record-keeping, including details of strikes and other action. Records were not kept by the trade unions either. Second, the high turnover of both managers and union officials meant that estimates of conflict levels are likely to be unreliable. Estimates are limited in accuracy, in any case, even under more favourable circumstances (Kelly and Nicholson, 1980). In the present case, estimates were obtained from the personnel manager and the two full-time officials (who had dealt regularly with the organization over a long period of time). Estimates were obtained (averaged across informants) for one year prior to the main impact of recession and for the last complete year prior to the study and are given in Table 7.3. Typical of much of the engineering industry, this was not a strike-free company prior to the onset of difficulties. The 1979 figure includes the national engineering strike but apart from this in both years strikes were of short duration (under three days) and involved only sections of the workforce. Leadership of strike action was in the hands of section stewards rather than senior stewards. Pay and discipline were key issues.

Industrial action short of strikes could cover, for example work to rule, withdrawal of co-operation, overtime ban and so on, but in this company it predominantly took the form of the overtime ban. With regular high levels of overtime it is not surprising that management was vulnerable to action in this area. A qualitative impression from a

Table 7.3 *Indicators of conflict*

	Strikes	Other action	Works conferences
1979	5	5	4
1984	8	15	15

range of stewards and managers was that lesser action had increased since the redundancies. Further, both managers and stewards asserted that the *threat* of an overtime ban was a powerful weapon for the unions. So the figures on lesser action probably belie its real significance as a power resource for the unions. While this had been true for 1979 as well as for 1984, both parties felt that management had become more vulnerable to the threat of an overtime ban recently because of the drive for production.

External works conferences are meetings to resolve disputes which require the involvement of the external officials of the unions and the employers' federation. They can be a useful indicator of conflict because they show that the parties are unable to reach a satisfactory outcome to an issue in dispute and must ask for assistance. This solution to disputes has some cost to both parties. The figures for 1979 and 1984 suggest a major increase in the use of this procedure. The union full-time officials and the employers' federation representative all believed that they were over-involved in the company's industrial relations because of the inability of the internal parties to resolve their own disputes satisfactorily. One union official joked that he planned to bill the company for bed and breakfast because he spent so much time at the plants.

The full-time union officials attributed their over-involvement partly to senior stewards' lack of skills: insufficient confidence in handling matters themselves and insufficient experience in deciding when to lose issues rather than take them through procedure. They also complained that the new company management frequently failed to follow procedure.

The frequency of works conferences frustrated both lay and external union officers. They had become such a regular feature of industrial relations in the company that most stewards had an *expectation* that disputes would not be resolved without a works conference. This expectation contributed to a sense of powerlessness, and dependency on the external officials. It is also possible to view works conferences as an indirect indicator of union sectionalism in that senior stewards were not seen to be sufficiently powerful to achieve settlements on their own but required external help.

Increasingly since the redundancies, works conferences had failed to resolve disputes. In 1984 nearly half the conferences resulted in a 'reference back' to domestic procedure. This meant, in effect, that the conference could not find or impose a solution because of entrenched positions. Here we see the prolonging of conflict: issues cannot be resolved internally and then with the passage of time are not resolved externally either. Issues, however, continued as a source of frustration and dissatisfaction.

The Power of Each Party

To examine power we must look at the objectives each party is trying to achieve and the power resources each holds. After this we can assess the power relations between unions and management, and make some judgements as to whether relations have been changed by the circumstances of decline and job insecurity.

Managerial Objectives and Power Resources

There was no written industrial-relations policy or set of objectives in the company and in this respect it was not unusual by British standards (Marginson et al., 1988). However, it was clear, from interviews and decisions made by management, that industrial relations were secondary to production and finance. Yet herein lay a major contradiction since the concern with production meant that management were dependent on the co-operation of the workforce.

Insofar as management had objectives concerning industrial relations, these were to hand responsibility for industrial relations from personnel to line management; to contain the influence of the unions; to introduce greater labour flexibility; and to gain control over pay levels.

Senior management wished to tie personnel matters more closely to production. This was to be achieved partly through changes to the payment system but also by making line management responsible for industrial relations. The latter was a popular development in the 1980s (Legge, 1988) but had extra significance in a company fighting for survival. Senior management were critical of the personnel department's approach to industrial relations, seeing it as dominated by elaborate and indulgent procedures more appropriate to a large, profitable company. However, without training for line management and without an overall industrial-relations policy, industrial relations became fragmented and inconsistent between plants and even sections. It reinforced the fragmentation and sectionalism already flourishing. Inconsistencies in pay, overtime and working arrangements became more marked. A number of disputes, and potential disputes, developed, which the personnel manager, in his now limited advisory capacity, did his best to defuse. The variation in action between line managers also exacerbated the tendency for more senior management to step in to deal with disputes which threatened production. This added further to the confusion and frustration not only of the shop stewards but also of the line managers.

The 'reassertion of control' was also concerned with containing the influence of the unions in the workplace. Management were successful in 1985 in limiting the time each senior steward could spend on union duties from full-time (informally) to eighteen hours a week. Although the unions formally protested, the decision was accepted de facto. Management also attempted, unsuccessfully, to rationalize the shop steward network by refusing to recognize certain stewards, by trying to reduce the number of stewards in certain areas and by trying to encourage an informal norm for constituency size. Stewards were also regularly reminded of the cost to the company of their monthly joint shop-steward meetings (generally two or three hours a month), which irritated stewards.

The desire to reduce union influence was not 'macho management' (Edwards, 1985c), in the sense of an aggressive drive to undermine union influence. Nor was it an example of the attempts, noted in some enterprises in recession, to bypass unions by using other channels of direct communication with the workforce (Marginson et al., 1988; Batstone and Gourlay, 1986). No alternative communication structures were established. It was a contradiction to attempt to reduce the scope and influence of the unions while failing to create alternative communication channels with the workforce, despite the need to do so at a time of great uncertainty and rapid change. An alternative strategy might have been to encourage steward development and training and to incorporate the unions into managerial decisions, procedures and, to some extent, values and perspectives.

A major objective for management was to reduce labour costs. The new vehicle was one means of achieving reductions through manufacturing design. But another way was to modify the payment systems and to control annual wage increases. There were several different payment systems in the company based on skill level and area of work. Pay consisted of a basic rate and a revenue-based group bonus, though both these varied from section to section. The new accent, popular in recession, was on 'ability to pay'. In addition, the company had introduced two years earlier the revenue-based bonuses. The one exception was a department in assembly which in spite of managerial pressure to change had retained a generous payment by results (PBR) scheme (which five years earlier had been company-wide).

Management's success in reducing labour costs was mixed. The annual wage rounds, an important element in predicting and controlling costs, dragged out in both 1984 and 1985 over several months, contrasting with the earlier years before recession. The delays caused a great deal of uncertainty to management and frustration to the workforce. The unions had presented their claim well in advance. In

1985 the wage claim was the subject of a works conference because the union had no response to its claim after four months. This strained union–management relations. Management succeeded in holding down basic rates of pay, which fell against the district average for the engineering industry between 1979 and 1985. The only major failure in this respect was the one section still on PBR despite strenuous efforts by the management to remove it. The opportunities for buying out PBR were limited given constrained finances.

While management controlled basic rates of pay, the dissatisfactions this generated among the workforce put pressure on the bonus and on overtime. The shop stewards felt that the bonus systems were, in principle, fair and reasonable. However, they were dissatisfied with the bonus in practice. Had production levels been higher, it might have operated satisfactorily, but often it did not reach the threshold level for payment. There was a series of small disputes as the unions tried to interpret the bonus agreement in ways to yield some reward. The bonus constituted a running sore which occupied much of the time of managers and stewards.

Furthermore, management did nothing about overtime levels, which remained high with stop–go production. It was an accepted practice month after month. Management believed that production and revenue would be jeopardized without it. But it meant that labour costs were not reduced and also that management were vulnerable to industrial action or its threat.

Management also wanted to increase labour flexibility and they had partially succeeded. As a result of job insecurity over the last three years workers had become willing to move to different sections on a temporary basis within the same plant and to undertake work which was different from their normal duties. Stewards and management dated this from the onset of redundancies. Although some labour flexibility was now in force, the situation had not moved to unilateral control by management. There were clear rules for relocation on a temporary basis and these were closely monitored by the shop stewards. Management were singularly less successful with changes in working arrangements. Attempts to move work rather than workers had been resisted on a number of occasions, especially in arrangements for the new vehicle. Management had backed down on some changes they wished to introduce. The changes rumoured for the new vehicle and the prevailing job insecurity had made stewards watchful and suspicious of change. This was not a climate for the rapid and smooth introduction of organizational changes or real flexibility.

Overall, when we consider the objectives of management and how far they were achieved, we can see that success was limited and even contradictory. For example, wage rates had been held down but wage costs were still high through overtime. Flexibility had been introduced but the workforce were more suspicious about change than previously. Organizational decline and the existence of job insecurity had not provided management with the power to deal with the workforce as they wished. The idea that job insecurity or fear of unemployment provides an opportunity for 'macho management' was not substantiated here. Indeed, from a management perspective, it was fortunate that the unions were not more informed or organized because the management were certainly not in a position to dictate terms.

The basis of managerial weakness lay in the nature of their dependency on the workforce. Dependency is critical to the analysis of power. In a context of financial stringency and crisis, management were highly vulnerable to shop-floor action or even the threat of action. Production, especially in daily changing circumstances, could not be achieved without the co-operation of the workforce. Thus, where the unions were motivated to gain a certain end and could organize to pursue that end, management had little option but to concede, because of the pressures on production. Also, daily firefighting had driven out longer-term planning, especially over industrial relations. Management were constantly reacting to rather than anticipating events. This increased their dependency on the workforce because they failed to anticipate the consequences of decisions. Management narrowed their focus to production and revenue and paid insufficient attention to issues of staffing and job security which were important to the workforce.

Union Objectives and Power Resources

A key aspect of worker power lies in trade union collectivity, and the ability of the union organization to express and resolve different interests within the membership. Yet we have seen that the unions were characterized by a high degree of sectionalism. How then did they pursue union goals and how successful were they?

The objective of strong union organization was elusive. Union organization had been weakened directly by the waves of redundancies and also by job insecurity which made workers less willing to become shop stewards and more vulnerable in the role. The company decline had also made the job harder as union representatives had to deal with a fragmented management. Informal workgroup pressures had also contributed to a lower union profile and the lack of continuity of stewards had increased problems of sectionalism. When management proposed and then implemented a limit on senior stewards' hours, the union organization was in no position to contest it. The restrictions, though, can be seen as a significant loss to the unions in their ability to organize effectively.

The unions were clear and insistent that they wished to maintain jobs in the company. Worker flexibility was accepted on a temporary basis. This was instituted with steward consultation. There was some success in preventing work being moved from one part of the organization to another. Considerable vigilance went into collecting information and analysing rumour about any aspect of management activity or decision-making with possible implications for jobs. However, the unions were not central to managerial decision-making and they were forced into the position of being reactive rather than negotiating changes in working arrangements. The company also employed some temporary labour and the unions had been unable to make these permanent.

On pay (an important issue to a workforce once nearly the highest paid in the district), the unions were frustrated in their attempts to achieve increases in basic rates. This generated a considerable level of dissatisfaction and a feeling of being taken for granted (or even taken for a ride) by management. There was a feeling that the workforce had been asked to make too many sacrifices over the last five years and these were still continuing for the sake of company survival. Given the scepticism about and lack of financial information there was the basis for considerable discontent over pay. Apart from the one section still on PBR, a chronic sense of grievance about pay existed in the company and there were several disputes or near disputes about comparisons within the company. The unions had also been unable to prevent the introduction of the revenue-based bonus scheme and there were several grievances about the bonus scheme. Several went to works conferences but were referred back to domestic procedure, which meant effectively that the unions had been unable to press their case.

The unions were potentially in a strong position on overtime. The threat to ban overtime was a powerful weapon. However, overtime had become institutionalized: in many ways it was as essential to the workforce as it was to management. It compensated for losses incurred through basic pay and the bonus. Thus, stewards would not ban it except where their members felt particularly strongly about an issue. It was not used to press wage claims because it functioned as a substitute for wage increases.

There were differences between sections as to how much overtime was worked and was seen to be acceptable. This was a source of union weakness and division since overtime dispensations created tension and suspicion between stewards. Distrust arose because some stewards felt the senior stewards were too zealous in policing overtime arrangements. Some stewards felt this was motivated by envy. Senior stewards were suspicious of collusion between stewards and managers in some departments to disguise the number of hours of overtime worked. Such distrust eroded a sense of collectivity in the union organization. Overtime was banned by sections to pursue local interests; it was never enforced across the whole company.

Had the unions been able to overcome their sectionalism the over-time weapon might have been stronger. As it was, workgroup was in competition with workgroup over the allocation of jobs, over differentials, over overtime and over the influence of shop stewards. There was little sense of unity. The unions were as fragmented as the management. Sectionalism made it difficult for senior stewards to provide leadership. The union organization, in fact, relied very heavily on external full-time officials for progressing any matters of substantial interest.

Union–Management Relations and Power

Overall, union organization had deteriorated since the onset of market and employment difficulties in the company. Feelings of job insecurity contributed to the lack of stability of the steward network. The rapid changes, especially in jobs, also contributed to a weakening of union power.

However, neither was management in a strong position. The focus on production to the exclusion of industrial relations had left management vulnerable to, and failing to anticipate, industrial action or the threat of action. Had the unions been stronger then management might have been even more vulnerable than it was. The sectionalism among the workforce prevented concerted action on issues.

We can see that both management and unions *won and lost* some issues of importance to them. It was not simply that job insecurity had weakened the unions to the advantage of management or that rapid changes in market forces had created pressures on management to the advantage of the unions. That would be a zero-sum view of power. The situation was more complicated: *each* party had been weakened by the straitened circumstances they were in. Vulnerability derived from their dependency on the other party. For management, worker co-operation was necessary to maintain production and hence revenue. For workers, change could herald job losses and so it was unwise to push claims. In any case, the fragmented union organization precluded the collectivity necessary for company-wide action.

The dependencies each had on the other had weakened each party and resulted in prolonged conflict. Industrial relations had become less satisfactory to both parties. Neither party was able to impose its objectives and so they limped along with high levels of tension, frustration, scepticism and accusations of bad faith.

Management did not have the resources of a more thriving organization to buy out discontent. Penny-pinching and cost-conscious, they made short-term decisions over labour which stored up longer-term difficulties. The situation of weakness for both unions and management meant that industrial relations, like the rest of the functioning of the company, was being run on a crisis basis, with little evidence of longer-term problem-solving.

Summary

In this case study we have had a rare opportunity to examine the twin effects of crisis and job insecurity. The events described here could happen to any company which suddenly experienced major and rapid changes, whether through restructuring, takeover or changes in markets or exchange rates. External difficulties revealed and magnified internal weaknesses in organizational functioning. Resources were limited so that certain options, open to more prosperous companies, were not feasible. A major restructuring of industrial relations along with the company was not possible. This should remind us that the dramatic economic changes of the 1980s do not automatically lead to 'new realism' (see also Kelly and Kelly, in press; Legge, 1988) or to 'macho management' or to any other type of change. In considering changes in industrial relations, there is a need to consider the financial and other resources, the organizational constraints, and the perceptions and goals of parties.

Job insecurity has been an important element of understanding industrial relations in the case study. Job insecurity was not uniform across the workforce: some workers were fearful for the continued existence of their jobs while others were cautiously optimistic and others simply did not know. The case highlights the shifting assessments of job insecurity with new events and rumours in the company. Any assessment of job insecurity by management or unions cannot be assumed to be static. The role of past experience on current understandings also requires further attention: at first sight it may seem strange that the level of apparent job insecurity is so low considering both the crisis management and the fact that the company was liquidated only eighteen months later. However, we also noted that those who expressed confidence in the future of their jobs saw current opportunities in the light of past experiences. At one stage the future of the company had looked so grim that by comparison current difficulties were viewed with some confidence. Furthermore, in this company, management failed to understand the significance of job insecurity, either for the workers themselves or for their own plans for developing the company. The management saw job insecurity as a peripheral issue but did not see its intimate relationship with trust, with resistance to change and with industrial relations.

Job insecurity was associated with a number of opinions and beliefs, such as trust in management and industrial-relations climate. However, we also saw some expectations confounded. In Chapter 6 we suggested that job insecurity might be related to union involvement and commitment but this turned out not to be the case. The degree of involvement in union affairs was not clearly related to insecurity. Participating in a union did not make workers less fearful nor does it seem that fear for job losses encouraged union involvement. Indeed, from interview data, we saw that the period of upheaval and uncertainty made workers somewhat less keen to be directly involved in union work and more cautious as shop stewards.

Job insecurity, also contrary to expectations suggested in Chapter 6, was not related to attributions of the causes of recent and current difficulties. It may be that these attributions were more socially developed than individually derived so that individual differences were not evident. The social nature of attributions is an area to which social psychologists are turning their attention (Hewstone, 1988, 1989; Moscovici, 1984).

One area where job insecurity played a key role was in the intertwining of insecurity, trust, industrial-relations climate and ongoing union–management relations. A significant relationship existed between insecurity and trust. Insecure workers trusted management less. This is an important finding. The low level of belief in the competence or intentions of management corroded a relationship already under strain because of high levels of uncertainty. The organization was undergoing rapid change and, in addition, the management had conspicuously failed to inform their workforce or their unions in any systematic manner about developments.

The issue of trust has surfaced repeatedly in this study and is known to be critical to good industrial relations (Fox, 1974; Purcell, 1981). Once trust has eroded it can be difficult to re-establish (Donaldson and Lynn, 1976). When low levels of trust are combined with the scarce resources of the company and the short-term, singlemindedness of the management's production objectives, then the prognosis for industrial relations is poor.

Furthermore, trust is critical to co-operation and adaptation. Job insecurity, combined with distrust, contributed to vigilance and a constant watchfulness about changes taking place in the organization. Although insecurity had contributed to greater worker flexibility in job tasks, it made other changes more difficult. This can be described as resistance to change. Greenhalgh and Sutton (Chapter 8) also note the relationship between insecurity and unwillingness to change. Yet change and adaptation are critical for a company in difficulties.

In Chapter 6, we suggested that power needs to be analysed not simply in terms of outcomes, but in terms of the objectives of each party and the extent to which they have achieved their goals. Various hypotheses about union–management power in the context of the fear of unemployment were proposed in Chapter 6. To see decline as contributing solely to macho management or to greater co-operation is unrealistic. We need to understand the particular organizational setting and the processes of union–management relations.

Here we saw that the unions were dependent on the management for the continuation of jobs. Their power base was also dissipated through the inability to reconcile sectional interest, which Batstone and Gourlay (1986) suggest is critical to a union power base. The management, for their part, were dependent on the unions for co-operation to maintain production and to make daily adjustments to the build programme. Conflict intensified, though both parties were weakened through their dependencies. Greenhalgh (1983) suggests that organizational decline may contribute to an intensification of conflict. Strauss (1984) suggested a similar outcome, arguing that whether or not conflict occurs will depend on the quality of the relationship between unions and management. The case here supports the view that increased conflict can occur although this does not preclude other outcomes in other settings. But it does suggest that any simplistic notion that the experience of crisis leads to greater co-operation through shared goals is misleading. Organizational psychologists have sometimes promoted such an approach to organizational problems (for example, Blake and Mouton, 1962) but it is clearly highly problematic (see also Kelly and Kelly, in press; Brown, 1988).

Overall, the case study has shown the role of job insecurity in understanding a situation of confusion, uncertainty and rapid change. However, job insecurity is insufficient on its own to understand industrial relations. It is important to understand the context of organizational functioning and the resources available. Job insecurity and other attitudinal and perceptual data such as trust and climate (but not apparently attributions) help in the piecing together of a complex pattern of organizational and industrial-relations processes. It is not enough to consider only the institutional level of analysis in industrial relations. Arguments solely at this level have failed to explain what has happened in recession and restructuring. Attitudinal data are essential if we are to understand not only the effects of recession, but change more generally.

8

Organizational Effectiveness and Job Insecurity

Leonard Greenhalgh and Robert Sutton

This chapter will show that job insecurity is both a cause and a consequence of organizational effectiveness. The chapter begins by defining organizational decline as diminished organizational effectiveness, especially in terms of the organization's ability to maintain the necessary level of adaptation to its environmental niche. We then examine how individuals working in that organization interpret such diminished adaptation, construe a state of decline, and react on the basis of how their work lives might be affected.

Workers' reactions have an impact on the organization's effectiveness because – in the aggregate – they determine the efficiency and adaptability of the workforce. The resulting changes in the organization's effectiveness are noticed by workers and cause further workforce reactions. The result is the emergence of a set of powerful positive-feedback loops that, if not managed, can accelerate the decline. This system-level view of the factors that link job insecurity and organizational effectiveness lays the groundwork for Chapter 9, which is concerned with how organizations can cope with these dynamics.

Organizational Effectiveness

Many scholars have defined organizational effectiveness in terms of the organization's interaction with its environment (for example, McKelvey, 1982; Pennings, 1975; Katz and Kahn, 1978; Seashore and Yuchtman, 1967). An implicit biological analogy is used in all such perspectives, and it is useful to pause to examine the implications of using the analogy as the basis of understanding organizational effectiveness.

Organizations are seen as being much like organisms. An organism is a living entity that relies on inputs from its environment to survive. The organism prospers when it is well adapted to its environment, and deteriorates if it is not; if the maladaptation continues, the organism dies. The idea is that organizations experience analogous fates. If environmental inputs are reduced or halted, organizational prosperity and survival are threatened. Therefore organizations that have become poorly adapted must make changes – that is, they must innovate. The changes (or 'variations') may be in technology, markets, resources, programmes for transforming those resources, or other aspects of organizational functioning. After evaluating the adaptive value of the changes, the organization selects and retains the positive changes, as do species (see Weick, 1979).

Innovation is a key process in determining the organization's ability to reverse decline. The way organizational theorists think about the innovation process is borrowed from the theory of biological evolution. Caution is warranted in adopting this point of view to illuminate organizational processes, for a number of reasons. First, evolutionary theory is a population-level theory, and yet in this instance is being applied to a single organization. Second, organizational actors rather than an impartial environment determine the subprocesses of variation, selection and retention, and these individuals' actions are likely to be guided by their idiosyncratic preferences, limited knowledge of environmental pressures and consequences, and predictions of future environmental states, rather than by immediate environmental press. Third, the part–whole relationships in living organisms are very different from those one finds in organizations. The survival of a living organism depends on the durability of its components. If the heart or brain functions poorly, then an organism will decline or die because (except in rare cases such as organ transplants) the life span of the organism cannot surpass the life span of its parts. At the same time, since the members of organizational systems are only partially included (Allport, 1933), the life span of an organization is not linked to the durability of its parts. An organization may live for thousands of years after the death of its original members, as in the case of the Catholic Church. In contrast, all of the members of an organization may be removed or quit – as in the cases of organizational death or massive job loss – without threatening their survival because they are members of other social systems, or can become members of other social systems. With these caveats in mind, let us proceed to define organizational decline as a reduction in organizational effectiveness, or more specifically, maladaptation to an environmental niche (Greenhalgh, 1983; Zammuto and Cameron, 1985).

Awareness of Decline

Cameron et al. (1988) identify two types of organizational maladaptation to an environmental niche. A *niche,* in this context, is the segment of an organization's external context with which it exchanges the resources, activities, technical knowledge and legitimation required to maintain the organization's survival. Borrowing from the socio-biological literature, they describe k-type maladaptation as the circumstance that occurs when a niche changes and the organization does not respond adequately to the change; conversely, r-type maladaptation occurs when the organization's performance erodes within a stable or improving niche. Thus, k-type decline is caused by environmental changes and r-type decline is caused by internal changes in the organization. In either event, organizational effectiveness decreases.

In simpler terms, an effective organization is one that performs well in meeting the needs of customers, investors, suppliers, regulators and the community it serves. When the organization begins to fail on these dimensions, it is in decline because it is no longer well adapted to meeting the needs of the organizations and people that support it. Two types of decline can be identified because either the environmental demands change and the organization fails to respond adequately, or the environment remains stable but the organization no longer performs satisfactorily.

Both types of decline eventually are perceived by workers, through several means including interactions with customers, publicly disclosed information about the company's performance, and news reports. Workers closely monitor this information, because it concerns an important aspect of their lives, and they discuss it among fellow workers to arrive at a shared assessment of the organization's level of adaptation. They realize that maladaptation affects the organization's prosperity and therefore it can subsequently affect them. As a result, they pay attention to what they see as evidence of the organization's ability to continue to maintain the size of its current workforce, to maintain current levels of compensation, and to maintain current levels of worker benefits.

Awareness that the organization is out of step with its environment is insufficient to provoke job insecurity. Poorly adapted organizations can and do survive, especially if they are large, old, diversified or enjoy monopoly status. It is only when workers see a decline in resources and believe that the organization is about to take action to deal with the maladaptation that they may perceive their jobs to be at risk. In this sense, the onset of job insecurity involves a transformation of beliefs about what is happening in the organization and in its environment. More specifically, workers become anxious due to the fear of adverse job changes, the most traumatic of which is job loss. In many cases, the event that gives rise to the perception of risk is the release of unfavourable financial data, particularly a drop in profit (or a budget reduction in a not-for-profit organization).

Greenhalgh and Rosenblatt (1984) define job insecurity as perceived powerlessness to maintain desired continuity in a threatened job situation. Thus defined, job insecurity has a spectrum of attitudinal and behavioural consequences. Some workers may respond by increasing job involvement and effort in order to save their threatened jobs. We provide some examples of such reactions below. But, more commonly, worker reactions stem from the tendency to withdraw psychologically from the job. They do this because the potential loss concerns a vital aspect of their lives. In this context, withdrawal in the face of possible loss protects the individual from the most severe sense of separation, and therefore is a basic ego-protective drive in people.

Job Insecurity Reactions

Chapter 3 presented a comprehensive discussion of reactions to job insecurity. This section covers some of the same phenomena, but builds towards a model of the *organization* as a dynamic system. In so doing, it presents a particular point of view based on theory involving social exchange and psychological withdrawal. The discussion is compatible with Chapter 3, but is narrower in scope, and focuses on the feedback loops that influence organizational effectiveness.

The reactions to job insecurity that will be highlighted in this section include (1) propensity to leave, (2) reduced job involvement, (3) diminished job effort, and (4) curtailed organizational commitment. The empirical evidence for each effect is briefly reviewed, then a model is presented that summarizes their combined impact on organizational effectiveness.

Propensity to Leave

When people perceive that their jobs are becoming insecure, they may think about leaving the organization (see the discussion of 'exit' in Chapter 3). The more valuable the worker, the greater is the chance that he or she will leave, with obvious harmful consequences for organizational effectiveness. These widely recognized conclusions are based on fragmented research evidence.

Chinoy (1955) reported a positive relationship between job insecurity and propensity to leave in his interviews with auto workers, but did not provide empirical data. Gow et al. (1974) found a positive correlation between satisfaction of need for security and tenure. Tenure can be construed as the inverse of turnover; thus these data are consistent with the thesis that many individuals are predisposed to leave an organization when their need for security goes unsatisfied.

The relationship was more explicitly studied in research by Greenhalgh (1979) and Jick (1979). The two studies used different measures of job insecurity and propensity to leave, and different samples of the same declining and shrinking hospital system. Both reported a significant negative correlation between the variables. (See also Ashford et al., 1989.) Additional research revealed that the more valuable workers, those who have better labour-market alternatives, are likely to be the first to leave (Greenhalgh and Jick, 1979;

Greenhalgh and McKersie, 1980; see also Levine, 1979; Whetten, 1980a). A similar pattern was observed in Sutton's case studies of eight dying organizations, in which he observed that, after the closing announcements, 'the best employees jump ship' (1983: 393). A set of macro-level studies corroborates the above findings concerning the relationship between job insecurity and propensity to leave. Cross-sectional analysis conducted by Fry (1973: 49) determined that industries with high involuntary job-loss rates 'also experience high quit rates since individuals will voluntarily move from companies or industries which possess a high degree of instability'. Fry's findings are generally supported in the research of Stoikow and Raimon (1968) and Block (1977).

An increase in propensity to leave does not, however, always lead to increased turnover. Propensity to leave will lead to turnover only when there are alternative opportunities (Price, 1977) and no strong dependencies (such as having skills that are uniquely needed by the current organization, needing to stay employed until one's pension matures, or facing disincentives for leaving such as deferred compensation or contractual penalties). An employee who no longer wishes to be an employee, but is constrained from leaving by either poor opportunities or dependencies, will substantially redefine the psychological contract (Rabinowitz and Hall, 1977; Schein, 1979). Furthermore, where severance payments are large – such as in the UK – workers are hesitant to quit because doing so means they will forfeit these considerable economic benefits.

The reluctant participant can be thought of as a 'psychological quit' (March and Simon, 1958; Quinn, 1973). That individual's degree of intended contribution would be sufficient to balance his or her perception of the exchange relationship, subject to a lower limit representing the minimum contribution necessary to avoid being fired. A worker's contribution is multifaceted, and can be differentiated into several categories: job involvement, job effort and organizational commitment are discussed below.

Job Involvement
Insecure workers are likely to reduce their level of involvement in their jobs, as another manifestation of psychological withdrawal. The concept of job involvement has been beset by theoretical and operational difficulties (Rabinowitz and Hall, 1977), and little research has been undertaken to explore the effect of a declining organizational context on the variable. Only two studies have explored the relationship, and neither presents a definitive test.

Hall and Mansfield (1971) used a shortened version of the Lodahl scale of job involvement (see Lodahl and Kejner, 1965) to measure the effects of organizational stresses on researchers in three research-and-development laboratories. Job insecurity was a key element of these stresses. Their hypothesis, based on balance theory, that the researchers would reduce their involvement in their work, was not empirically confirmed. Greenhalgh (1979) also used a shortened version of the Lodahl scale, but correlated job involvement with a three-item job-insecurity scale in a study of a declining hospital system. As hypothesized, he found a negative correlation between job insecurity and job involvement, but the magnitude was low and failed to reach significance.

Also relevant is the study by Gannon et al. (1973). They compared two groups of defence-industry engineers that had worked for the same company. The first set had been laid off and later re-employed, while the second set had never been laid off. The two groups showed no significant differences on any of the twelve Lodahl scale items used, but job insecurity was not measured. Instead, Gannon and his colleagues (1973: 331) assumed that because 'both the managers and their engineers knew that reductions in force were highly improbable . . . the current engineers should not be influenced in any appreciable manner by the reduction in force while the terminated engineers should be attitudinally affected in a negative manner'. Their implication is that the survivors constituted an appropriate control group. However, the assumption of high job security of the survivors is questionable (Greenhalgh, 1979, 1982; see also Chapter 2 of this volume), since individuals' assessments of job security seem to be based on a constructed reality that can be in direct contradiction to 'objective' reality. Thus it is quite likely that both of the groups studied by Gannon and his colleagues were equally insecure; those who left, because of the direct experience of job loss, and those who survived, because of the vicarious experience of job loss. As a result, the study must be viewed as inconclusive.

These sets of results, obtained using less than ideal measures of the variables, lead one to question whether job involvement is an aspect of the psychological contract and therefore vulnerable to adjustments of exchange imbalances. Hall and Mansfield (1971) found that individuals' scores on the scale were fairly stable compared to the variation across subjects; they suggested that the construct may represent more a personality trait than an attitude, a possibility raised in the original Lodahl and Kejner (1965) article, and in several papers since (Lawler and Hall, 1970; Rabinowitz and Hall, 1977; Schwartz, 1980).

Job Effort

Job effort is the most basic contribution in an exchange relationship. When the declining organization is viewed as withdrawing continuity of employment, one hypothesis is that workers will respond by putting in less effort to restore the balance of the exchange. An alternative hypothesis is that workers will respond by putting forth more effort in an attempt to save their jobs, or their organization. The empirical evidence linking job insecurity to job effort is mixed, but greater support is found for the first hypothesis that workers respond by decreasing their efforts.

One prevalent reaction to potential job loss, particularly when the worker first hears the rumours or news, is anxiety. Several articles have been written on the relationship of anxiety to productivity, mostly involving laboratory experiments (see, for example, Taylor and Spence, 1952); they indicate a general impairment of performance associated with high levels of anxiety. More relevant to field settings is the work of Roethlisberger and Dickson (1946: 153), who attributed to job insecurity anxiety some aberrations in their data from the Hawthorne plant:

> The . . . decline in output . . . was . . . related to the operators' anxieties over the uncertain future of the mica splitting job. The decline began shortly after the first rumors . . . appeared, and it progressed as . . . fears . . . became more acute. This experience showed the effect of interfering pre-occupations on the attitudes of the operators and, in turn, on their output.

These findings do not constitute rigorous support for the hypothesized positive relationship between job insecurity and reduced productivity, since job insecurity was inferred rather than measured. If 'hard' measurements rather than observations had been made, however, the results would have been very useful, since the quasi-experimental design, even though applied fortuitously to a natural field experiment, had some real strengths. The researchers made pre-, during and post-observations on both the experimental population and a matched control population. The effect was absent in the control and pre-treatment observations, and attenuated in the post-treatment observations. Thus the Hawthorne findings appear to support a relationship between job insecurity and diminished work effort.

More direct support was found by Greenhalgh (1979), who found a significant negative correlation between job insecurity and effort. Both variables involved self-reports; the effort variable encompassed current work effort and behavioural intentions concerning effort. Hanlon's (1979) interviews of workers who lost their jobs in the New York City fiscal crisis corroborated these findings. Both studies involved US public sector employees who survived a lay-off of their co-workers. Because these survivors had believed their jobs to be invulnerable to cutbacks, they were shocked when other employees lost their jobs. The survivors spent many of their work hours either incapacitated by their anxiety or talking with other workers about the likelihood of further cuts. Their productivity plummeted, but productivity among public servants is often hard to measure directly (unlike a factory, in which the decrease in output is immediately obvious), and therefore these workers were not punished for their drop in performance.

Restriction of output is also related to job insecurity. Guest and Fatchett (1974) noted that a threat to job security often resulted in a conscious decision by workers to restrict their productivity, with the intention of alleviating the threat by 'stretching out the job'. Beynon (1973) found evidence of the relationship between job insecurity and diminished productivity in his observations over a five-year period in the British motor industry.

A pair of additional studies which bore directly on the hypothesis found no relationship between job insecurity and productivity, while another stream of research suggests conditions under which positive correlation may be observed. Hershey (1972) compared piece rates of matched groups in the same organizations; the 'experimental group' (this was a natural experiment) had been notified of their job loss, while the control group had not. All four companies where the research was conducted had histories of job loss, and control subjects were chosen on the basis of their being vulnerable to job loss. Thus, Hershey's data do not refute the present hypothesis, since Greenhalgh (1979) shows that there should be little difference between the threat (that is, the vicarious experience) experienced by the control group and the threat experienced by the experimental group. Hershey's research would only represent a test of the hypothesized relationship between job insecurity and productivity had he considered his two groups alternative treatment groups, and compared these with matched control groups consisting of non-vulnerable workers in companies with a history of perfect job security. Second, Hall and Mansfield (1971) studied reaction to job security stress among researchers in R&D laboratories and found 'no significant differences in self-rated performance or effort' between their affected and control groups.

Brockner and his colleagues have conducted an innovative series of laboratory studies in the USA that are relevant to this topic (see, for example, Brockner et al., 1985). They simulated job loss among college student subjects by publicly laying off one of the participants (actually a confederate) in a short (eight-minute), low-compensation (five dollars) proofreading task. This paradigm permitted systematic assessment of the reactions of 'survivors', whose jobs were put at risk by the experimental manipulation. Their findings suggest that productivity tends to increase among subjects who 'survive' the simulated job loss, especially subjects who have low self-esteem.

Despite the gains in experimental control, this approach has severe shortcomings in terms of its external validity, as Brockner has noted. A college student's fear of lay-off from a simple, short-term proofreading task does not approximate the experience of workers who fear losing longer-term jobs in which their economic security, adult careers and identity are at stake. Thus, it is not clear that these researchers' initial findings that survivors increased their work output (see, for example, Brockner et al., 1986) can be generalized to real-world organizations in which jobs are at risk. Indeed, when these researchers extended their work to field settings, they actually found some evidence of decreased self-reported effort associated with the perceived injustice of the co-workers' job loss (Hurley et al., 1988).

Sutton's (1987) case studies of dying organizations also suggest that impending loss of a role can lead to increased effort and productivity under some conditions. His findings from eight case studies of how 'dying' organizations make the transition to 'death' indicated that, despite managers' fears that the announcement of the closing would cause the effort of lower-level workers to plummet, the announcements were typically followed by consistent, or more often increased, effort and work quality. A similar pattern of results was reported in Slote's (1969) case study of the termination of a paint plant, as well as in Kissler's (1987) account of the closing of ten General Electric plants.

Nonetheless, the finding that announcements that an organization will close are often followed by increased productivity does not contradict findings that job insecurity leads to reduced productivity. These studies suggest that the possibility that job loss may occur is a threat that causes members of declining and dying organizations to put less effort into their work roles. Once the announcement has been made that a closing is certain to occur, workers know that they are going to lose their roles. Sutton's (1984) case studies suggest that one reason that productivity increased after the announced organizational death was that workers had been paralysed by their job insecurity. The announcements of the closings were, of course, construed as bad news, but the official word removed the stress of uncertainty (see Chapter 2 for a discussion of 'event uncertainty' and Chapter 9 for a discussion of the important role of communication programmes in managing job insecurity). Since workers knew – rather than worried – that loss of role was going to occur, they could begin to take actions to cope with the impending loss.

Indeed, Sutton (1987) proposes that workers who know – rather than fear – that they are going to lose their work roles have strong incentives for working hard. Workers' performance in the dying organization constitutes their last chance to make a good impression on leaders who can recommend them for future employment. Workers in effect feel pressure to work themselves out of one job and into another. Furthermore, as the spectre of unemployment looms larger, workers are likely to become concerned with building up a cushion of savings, or at least minimizing their debts, so as to survive possible lean times ahead. In short, the literature on organizational death implies that job insecurity reduces effort among people who fear the possibility of losing their jobs, but not among those who are certain they they will experience job loss.

In summary, there seems to be a relationship between job insecurity and diminished work effort. It can result from such diverse factors as becoming incapacitated by anxiety, being distracted from the task by the need to discuss further threats with other workers, withdrawing effort due to the perceived abrogation of the psychological contract, stretching out the work to increase the employment period. Under some circumstances, this general tendency will not manifest itself, and when threatened workers have become resigned to rather than insecure about their fate, they may put in a short period of higher than normal effort. In general, however, we should expect job insecurity to lead to lower productivity.

Organizational Commitment

Job involvement and job effort are worker contributions which concern the work role, whereas the referent of organizational commitment is the organization. This construct has much in common with Hirschman's (1970) concept of 'loyalty', discussed in detail in Chapter 4. In a declining and shrinking organization, one would expect workers to adjust their psychological contracts (see Chapter 2) by reducing their commitment to the organization (see Hanlon, 1979: 105).

The concept of organizational commitment has long been recognized as important in the study of organizations, although much confusion exists as to what comprises the construct (Becker, 1960; Buchanan, 1974). Our conceptualization of organizational commitment, based on exchange theory, has its roots in the work of Chester Barnard, who viewed his roughly equivalent concept, loyalty, as 'an essential condition of organization' (1938: 84). In a similar vein, Porter et al. (1974: 608) pointed out that 'the development of organizational commitment appears to require an individual to think in fairly global terms about his or her relationship to the organization'. Consistent with Barnard, and Porter and his colleagues, the concept of 'cooperative' behaviour of workers involves their going beyond the content of supervisory instructions, job descriptions and standard operating procedures to feeling committed to the organization's mission. Thus, loyal workers who have decided to contribute high loyalty would be expected to be willing to participate in the give and take of an organization's contingent reactions to its environment: in other words, they would tolerate some degree of organizational change since they would be identified with its general goals.

From this line of reasoning, there emerge two collateral dimensions to organizational commitment: organizational identification and resistance to change. The distinction between the organizational-identification and resistance-to-change dimensions is more obvious in operational terms than it is in conceptual terms. In the case of identification, one in effect asks directly whether respondents feel a sense of organizational belonging (that is, with the organization, its management and the workgroup). In the case of resistance to change, one asks whether changes which may not maximize respondents' personal interests are viewed favourably. A highly committed worker identifies strongly with the organization and does not resist adaptive changes. This section of the chapter examines the extent to which these attitudes are affected by job insecurity.

Brockner and his colleagues (1987) surveyed workers in a 300-store sample of a retail chain that was in decline and had experienced widespread store closings with job loss. Using a multifaceted measure of organizational commitment, they asked respondents to assess their own attitudes both before and after the job losses they survived. They found a drop in organizational commitment, especially among those who strongly identified with the people who had lost their jobs, and attributed the effect to perceived injustice.

Only one study was found which investigated the specific relationship between job insecurity and organizational identification. Greenhalgh (1979) found a significant negative correlation, confirming his hypothesis that job insecurity engenders psychological withdrawal in situations of organizational decline. (See also Ashford et al., 1989.)

Resistance to Change
The relationship between organizational decline and resistance to change is conceptually complex because one would expect to find three factors that can work simultaneously to overdetermine the relationship. First, organizational decline leads to organizational change, and workers who perceive themselves to be affected by it could be expected to resist the change per se. This prediction is based on the contention of Fink et al. (1971) that perceived changes mobilize maintenance forces within the psyche that arise to re-establish homeostasis.

Second, there should be a simple direct relationship between decline-induced change, as perceived by workers, and their resistance to the change. March and Simon (1958) contend that resistance will occur if the changes reduce the surplus of inducements over contributions in the workers' assessment of the exchange relationship. In other words, resistance can be expected when workers feel that the organization appears to be asking too much of them in the light of what it is giving in return. Resistance represents an effort to hold the organization to the terms of the psychological contract.

Finally, as noted earlier, decline increases stress, which in turn makes organizational actors less adaptive. The threat-rigidity thesis (Staw et al., 1981) suggests that the threat of decline will cause both managers and workers to cling strongly to standard operating procedures when the organization most needs to change them. Fox and Staw (1979) have confirmed the existence of this 'trapped administrator effect' in a laboratory study, finding that as job security decreased there was a significant main effect on subjects' commitment to a previously chosen course of action. Similarly, the D'Aunno and Sutton (1988) study of 156 drug-abuse treatment organizations in the USA indicated that loss of organizational funds and funding sources was associated with more bureaucratic 'red tape' and more standardized jobs, both of which indicate a tendency to rely more heavily on established procedures.

These distinctions obviously have greater conceptual than operational clarity. The same recalcitrant attitudes and behaviours can be the result of each or all of the above factors. Nonetheless, researchers studying declining organizations can expect: (1) some resistance to any change, (2) conscious resistance to the specific changes that are altering the exchange relationship.

There is substantial literature on resistance to change that supports the hypothesized relationship between organizational changes, psychological contracts and worker resistance to change. Much of this literature involves the broad spectrum of technological changes (for example, Coch and French, 1948), although other studies involved reactions to reorganizations (for example, Rothman et al., 1971). Merton (1962) theorized a positive relationship between job insecurity and resistance to change. Greenhalgh (1979) measured the relationship specifically and found a significant positive correlation between the variables.

A Summary Model

Workers construct a definition of their employment situation, which includes a subjective assessment of the level of job insecurity (see Chapters 2 and 3). The workers' perceived risk of job loss is magnified by managerial secrecy, rumours, ambiguity and the rhetoric of interest groups. A secure job is an organizational inducement (March and Simon, 1958); therefore, the perception of diminished security may be viewed as a violation of the psychological contract and a resulting imbalance of the exchange relationship (Mowday et al., 1982). Many workers restore the imbalance by reducing their willingness to continue participation, or by reducing their contributions if exit is constrained. The correlates of job insecurity therefore include propensity to leave, job involvement, job effort and organizational commitment. Those dynamics have organization-level consequences, primarily through their effects on efficiency and innovation. Thus the psychological reactions of workers are shaped by powerful positive-feedback loops.

The psychological reaction of workers who actually lose their jobs

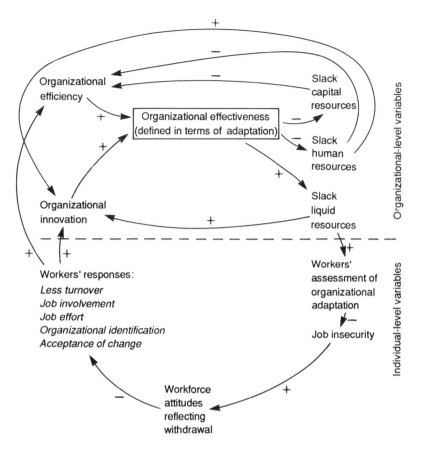

Figure 8.1 *Summary of organizational and individual-level variables involved organizational effectiveness*

also has organizational consequences: the experience leaves attitudinal scars, and most job losers subsequently obtain jobs in the same or other organizations. Employees who have suffered involuntary job loss in the past may be less loyal to subsequent employers because they have learned that the employment relationship is a temporary one that can be ended with little or no advance warning. The actual job-loss experience produces a qualitatively different type of psychological reaction than the vicarious experience (that is, job insecurity; see discussion in Chapter 2). The effects of actual job loss are better explained in terms of the grief paradigm than of exchange theory (see Greenhalgh, 1979). Indeed, workers who lose jobs as a result of organizational death may hold 'organizational funerals' (Albert, 1984) or 'parting ceremonies' (Harris and Sutton, 1986) so that they can mourn the loss of their work roles and their organization.

Figure 8.1 portrays the combined effects of these manifestations of withdrawal, and shows their impact on organizational effectiveness. Two features of the model deserve particular attention. First, maintaining a high level of adaptation requires both innovation and efficiency. Innovation is essential to organizational effectiveness if there is any environmental flux. When the niche changes, organizations must vary their technologies (including their programmes, human and non-human capital, and resources) and then select and maintain organizational properties that increase adaptation to their new environment (see Greenhalgh, 1983: 241). Failure to do so, as noted earlier, results in k-type maladaptation (Cameron et al., 1988). An effective organization, by the definition posited in this chapter, maintains its level of adaptation despite environmental fluctuation.

Efficiency is also important. The organization's processes must yield satisfactory outputs for the inputs it consumes, otherwise it will be denied those inputs (Katz and Kahn, 1978). Denial of inputs can take several forms: buyers will no longer pay the purchase price, investors will no longer invest, creditors will no longer give credit, patrons will no longer support the organization, or voters will no longer demand the programmes. As inputs are withdrawn, the inefficient organization will become ineffective and decline.

Second, the model involves positive feedback loops that, if unchecked, will tend to accelerate organizational decline. The positive feedback loops are an immediate consequence of changes in slack resources, which tend to be volatile indicators of organizational effectiveness. A detailed discussion of the effects of slack is provided in the section on buffering in Chapter 9. Here we will simply note that organizational decline differentially impacts capital, human and liquid resources.

Capital resources (such as plant and equipment) become surplus when the maladapted organization can no longer support its former scale of operations. They cannot be instantly disposed of, however, therefore they become an economic burden and consequently reduce efficiency. This relationship forms a positive feedback loop because the resulting inefficiency further impairs organizational effectiveness, which in turn calls for further reductions in the scale of operations, this generates more slack capital resources, and so on.

Slack human resources have a similar positive feedback effect because they, too, reduce organizational efficiency. Unlike the case of slack capital resources, organizational managers have some choice in the speed at which human-resource levels are reduced, a point that is emphasized in Chapter 9. Organizational slack is also involved in a possible negative feedback loop which operates through organizational innovation. As noted earlier in this chapter, innovation means 'doing something different' (rather than doing the same old thing more efficiently) in response to the niche demands, and involves the sub-processes of variation (that is, experimentation), selection (that is, choosing the more promising of the options that were introduced on a trial basis), and retention (that is, institutionalizing the preferred option). Each subprocess requires levels of human-resource inputs above steady-state requirements. Thus, the silver lining in the dark cloud of organizational decline is that the surplus human resources tend to be available at the time they are most needed for adaptive organizational changes.

This silver lining has some tarnish, however, if we may extend this metaphor further. Even though the surplus human resources are available that could be used to bring about adaptive changes, managers are unlikely to take such advantage of these workers' availability. There are two basic reasons for such inaction.

First, the stress of a decline situation seems to engender a myopic commitment to the existing way of doing things (Staw et al., 1981). Whetten (1979) explains this process occurring because organizational decline increases the pressure of accountability, particularly in US organizations. In response to the increased monitoring and perfor-mance-evaluation, decision-makers are likely to prefer to make their personal effectiveness apparent. They will therefore tend to favour courses of action that are highly visible and have consequences that are measurable in the short term. As a result, the selection subprocess may be shaped in a way that favours immediate efficiency and inhibits innovation.

For example, leaders typically focus on eliminating the workforce oversupply, which is a symptom of decline, rather than focusing on reversing the causes of decline. Despite such tendencies, firms may benefit if they use surplus workers to reverse decline through means such as developing new products or implementing new marketing strategies. Other farsighted leaders may use periods of decline to teach workers new skills, which will be valuable to the firm when demand for their labour increases. But such adaptive responses appear to be uncommon.

Second, the stress of decline adversely affects managers' decision-making behaviour in a way that goes beyond the rationality of account-ability pressures. Decline is threatening, and the threat produces a general tendency toward rigidity (Staw et al., 1981). In addition, managers are likely to prefer a quick response to relieve the tension they experience (Whetten, 1980b), which may preclude the thoughtful problem statement and diagnosis that is essential in addressing complex organizational problems. As a result, definition of the problem situa-tion as *critical* (that is, the focus is on immediate response) can lead to neglect of its *crucial* elements (those that are vital in importance irrespective of their urgency). Finally, stress tends to foster problemistic search (Smart and Vertinsky, 1977) whereby solutions are sought in the vicinity of symptoms and less obvious options are not considered (Cyert and March, 1963).

An example of problemistic search occurred recently when a US hospital was declining because of a shortage of nurses, and had to turn away revenue-producing patients that it badly needed to cover the large fixed costs. Highly stressed managers focused on recruiting, seeking ways to draw on the European supply. This was an example of 'searching for solutions in the vicinity of the problem' in the sense that the problem was defined in terms of supply. They had overlooked options such as trying to cut the turnover rate, redesigning jobs so that unskilled workers did work that didn't require nursing skills, and tapping the pool of former nurses in the area who didn't want to work full time, but were delighted with part-time work.

As a result of these dynamics affecting managers' responses, declining organizations tend not to take full advantage of the serendipitous surplus in human resources. They become preoccupied with maximizing efficiency and reducing the scale of operations per se, and don't consider how the available workers could be used to make organizational changes. Thus, they pass up an opportunity to increase innovation and thereby counter maladaptation.

The most important positive feedback loop, in terms of the present discussion, begins with the curtailment of slack liquid resources (Cameron et al., 1988), especially the amount of funds available to the organization. Financial data constitute the most volatile index of maladaptation and therefore provide an early warning of jobs at risk.

The organization-level effect of diminished liquid slack is to preclude innovation. Organizational changes consume slack resources, especially during the transition period. Even if no capital investment in new technology is needed, new routines involve a learning-curve whereby human resources must be used inefficiently during the transition period. The need to fully compensate these relatively unproductive workers consumes liquid slack.

The consequence of diminished liquid slack, like all of the effects discussed thus far, is a deterioration of organizational effectiveness. The drying-up of liquid slack is an anxiety-producing cue to the workforce that produces the chain of individual-level effects depicted in the bottom half of Figure 8.1. The individual level phenomena – job insecurity, withdrawal-related attitudes and corresponding behaviours – subsequently affect the organization-level phenomena of innovation and efficiency.

This positive feedback loop is particularly problematic because it cannot be managed directly by means of managerial decisions. It involves sense-making by workers, a mixture of individual and collective interpretation of daily events and forecasting of the consequences for continuity of employment. The next chapter discusses organizational coping strategies that can be employed in an attempt to counteract the adverse – often disastrous – effects of the positive feedback loop involving workforce attitudes, but it should be emphasized here that managers face an exceedingly difficult task in preventing the accelerative effect and an even more formidable task in reversing the damage once this particular causal loop has been activated.

The Worker's Predicament

The key factors involved in workers' experience of jobs at risk are prediction, understanding and control. Building on a large body of psychological research, Sutton and Kahn (1987) have proposed that, as human beings live their lives, they strive to anticipate the events that they will encounter (prediction), to know the reasons that events occur (understanding), and to influence the outcomes of events (control). The absence of these cognitive factors may produce stress, while providing these factors in an organizational-decline situation relieves stress. Thus, understanding how the workers experience their predicament is central to managers' decisions concerning organizational coping strategies, elaborated in Chapter 9.

Prediction involves being able to envisage possible outcomes and attach probabilities (or at least implicit probabilities) to each scenario. As has been pointed out in Chapter 2, the degree of uncertainty is independent of the valence of the outcome, and uncertainty can be more stressful than negative consequences. For example, a worker can cope with a certain loss of a job through anticipatory grieving (Greenhalgh, 1979). In such cases, uncertainty is low and negative valence is high. Often more stressful is the vague possibility of demotion or reassignment to an unpleasant task. In such cases, uncertainty is high and negative valence is relatively low. What makes the latter scenario stressful is the difficulty in coping with it. The worker cannot grieve the loss because the uncertainty provides a strong rational basis for denial of the loss, a stage in grieving that precludes coming to terms with the loss (Kubler-Ross, 1969, and Chapter 2 in this volume).

Understanding is essential for stress reduction because the worker needs to adjust cognitively to the predicament. For example, the subjective reality of risk is likely to create strong cognitive dissonance. The worker's belief that he or she is a valuable human being is incompatible with the possibility that the organization would want to get rid of him or her. If that same individual is given to understand that the organization is randomly or impartially (such as by inverse seniority) laying off even its good workers due to economic difficulties, then the dissonance can abate. Workers' malaise may be the result of perceived injustice rather than cognitive dissonance (Brockner et al., 1987), requiring a concomitant new understanding to alleviate stress.

Another reason why understanding is important is workers' attributions. There is a tendency, at least in the United States, for workers to believe that if an organization is in decline, it is somehow due to insufficient work effort. The consequence of this belief is that workforce reductions are believed to be carried out on the basis of merit, even if there is an outward appearance of an impartial seniority system operating. Accordingly, they believe that organizational effectiveness and the risk to their own individual jobs can be reversed by harder work or higher quality. This prevalent belief system can somewhat counteract the more potent tendency for insecure workers to reduce job effort, but it reinforces the coping mechanism of denial, and has the undesirable effect of reducing the surplus workers' motivation to find replacement employment. This increases the trauma of job loss and makes it more difficult for managers to reduce the workforce through attrition programmes (see Chapter 9 for details).

For these reasons, it is important for managers to take an active role in managing workers' understanding of the organizational decline situation. In the absence of such efforts, rumours, misunderstandings and belief tendencies are likely to create dysfunctional sense-making.

A sense of control is strongly desired by most people in any organizational change situation. Indeed, much psychological research indicates that human beings seek a dependent relationship between their actions and important events in their lives (Seligman, 1975). Similarly, Adler stated in 1930 that control over one's relevant environment is 'an intrinsic necessity of life itself' (cited in Langer and Rodin, 1976: 398). In the specific case of job insecurity, perceived lack of control obviously increases the sense of vulnerability, but it also generates frustration, which can give rise to an aggressive reaction that can exacerbate the withdrawal-related resistance to change of insecure workers.

Workers are likely to be denied control at the time they most need to exert it because managers tend to seek less input from subordinates under the threat of decline conditions (Whetten 1980a, b; see also Staw et al., 1981). Some of this effect is attributable to a threat-rigidity response, as discussed earlier. It can also be the result of a realistic assessment of subordinates' biases. Levine (1979) points out that while managers typically are aware that subordinate participation in decision-making facilitates change, they also are aware that subordinates are unlikely to help make decisions that may be best for the organization but are not in their own best interests.

In sum, there are several reasons why workers' stress levels may be influenced by the degree of prediction, understanding and control they experience. There are also several factors that determine managers' ability and willingness to provide these experiences. The result of these dynamics is a formidable challenge to any manager who wishes to minimize the workforce trauma that can result from a decline in organizational effectiveness. Such efforts by managers are the focus of the next chapter, which explores organizational coping strategies.

Workplace Effects

Before moving to Chapter 9, we should examine what happens to the workplace as a whole. When an organization is in decline and workforce shrinkage is planned or has begun, the climate of the organization changes. The climatic changes are more than the sum of individual reactions. To illustrate, let us consider what happens to the typical workplace when it undergoes a transition from stability to decline.

When the organization is enjoying a stable phase, the focus of workers' attention is on task and relationships. Workplace conversations tend to reflect this focus. People talk about the work itself, and about what it is like to work with other people. The conversations reflect dominant workplace values. For example, in a stable organizational situation in which workers are highly committed to the task, they will exchange stories of task mastery and task inhibition (factors – including people – that get in the way of doing the job the way it ought to be done). Commonly recognized heroes will be the people who respond successfully to task challenges, co-operate with others, or embody the idealized company image. In a stable organizational situation in which workers are alienated, conversations will emphasize the unreasonableness or ineptitude of managers, or the exploitation of workers. Commonly recognized heroes will be workers who defy overbearing managers, denounce managerial incompetence, get away with exploiting the company or leave the company for greener pastures. A large part of the focus of attention will be on the union or works council, and workers' outside lives.

When the organization moves into a decline phase, workers quickly recognize that the organization is becoming maladapted and surmise that shrinkage is in the offing. The climate of the workplace changes along several dimensions. The collective consciousness temporarily disappears, and is initially replaced by individual preoccupation with how the shrinkage will affect each person. Workers interact with each other, primarily discussing what might happen to the workforce, but the character of the interaction changes. In the stable phase, much of the interaction resembled a ritualistic communication that preserved a shared reality. In the decline phase, it becomes transformed to a more desperate, almost obsessive preoccupation with one's own continuity of employment and economic well-being.

In time, the focus returns to a sense of shared fate, but this time attention (that is, the high interest and emotional energy) is on continuity and upheaval rather than on task or relationships. Inter-personal bonds weaken, as workers become acutely aware of their zero-sum interdependence: when positions are scarce, a co-worker's survival in the organization might mean one's own demise. Thus trust diminishes, and people become more guarded in what they share and disclose, which further erodes trust, and changes the nature of interpersonal relationships.

Workers' relationships with the organization change along with the transformation of relationships with co-workers. Workers' connectedness to the organization varies along a continuum from low to high involvement and sense of belonging. When the workforce experiences the job insecurity crisis, the connectedness tends to sink to the low end of the continuum, become purely calculative (Etzioni, 1961), and instrumental norms ('we do the job only because we get paid for it') pervade organizational culture. Thus, whatever the shared reality beforehand, when the organization is in decline, people tend to become concerned with job rights and organizational entitlements. The workforce therefore displays limited organizational commitment. Their shared definition of the situation characterizes the organization as showing little commitment to its workers, and therefore deserves only the minimum worker contribution in exchange.

Managers as a group construct their own shared reality. They, too, experience job insecurity, but usually not with the same intensity that is felt by lower-level workers. There are several reasons for this differential experience. First, they realize that job loss almost invariably affects lower-level workers first, so they expect plenty of advance notice before their jobs are really at risk. Second, in many cases they are more mobile in the labour market, having generic managerial skills that are seen as useful to alternative employers. Third, they tend to have amassed more resources (for example, savings, assets that can be liquidated in an emergency, and borrowing capability) with which to cope with adverse fortunes than do their lower-paid counterparts. Finally, they are in power positions due to their organizational roles and do not experience the same degree of abject powerlessness that is felt by lower-level workers.

This is not to imply that managers are unaffected by organizational decline. They do, indeed, experience job insecurity, even though it tends to be less traumatic than that of lower-level workers. They also may experience intense stress and guilt as they implement shrinkage decisions. The process of workforce reduction may involve the wilful act of firing people the managers know well, people with whom they have social and interpersonal relationships that go beyond the scope of work roles. The effects of their decisions are palpable: the managers can empathize – and perhaps even sympathize – with the job losers, understanding and even sharing the pain that their decisions inflict.

Managers need to cope with these reactions to their workforce-reduction decisions. Coping involves some form of emotional insulation. Defence mechanisms such as rationalization are common: 'It's not really my decision that's causing such pain; the whole industry is letting people go, so if it wasn't me letting them go, it'd be someone else. These people just picked the wrong industry to work in.' This defence mechanism is adaptive, because it keeps the manager from being overwhelmed with emotions, allowing him or her to keep functioning effectively and to be humanistic towards the affected worker.

Less hospitable is the defence mechanism of depersonalization: 'These workers are just a commodity, and they should realize it. They know the laws of supply and demand; they shouldn't expect to keep their jobs if the company suffers a downturn.' One of the worst is depersonalization with machismo overtones: 'The manager's job calls for tough decisions, and one should be man enough to make the tough decisions even when people get hurt. This is a business we're running, not a charity.' Unfortunately, across the world, machismo values are not uncommon in many organizations facing workforce-shrinkage decisions, and they leave little room for humanistic accommodation.

The important point to note is that managers' coping reactions can do a lot to shape workers' experiences of the downturn, with the general effect being the less humanistic the approach, the more adverse the change in organizational climate within the surviving entity.

This brief account of how managers might act in situations which call for them to make decisions about how to deal with job insecurity paves the way for Chapter 9, in which organizational coping strategies are discussed.

9

Organizational Coping Strategies

Leonard Greenhalgh

Influencing job insecurity requires a balancing of what managers need to do to adjust the workforce when shrinkage is necessary, and sensitivity to workers' experience of jobs at risk. Specifically, the stress that exacerbates adverse reactions is a function of the extent to which workers experience prediction, understanding and control of the risk to their jobs. Obviously, managers can play a powerful role in shaping worker sense-making. But in addition to influencing subjective phenomena such as perceptions, understandings and emotional responses, managers also must influence objective phenomena, making adjustments that will restore the organization to an adequate level of adaptation.

Shrinkage is not always necessary, however. Enlightened managers will understand that diminished organizational effectiveness results from poor adaptation of the organization to its environment. Such insight gives them other options for coping with a decline situation. They can make organizational adjustments to restore the level of adaptation, rather than reduce the level of operations commensurate with what the poorly adapted state will support.

Adaptive adjustments can be made in many areas. Managers can move into new markets, introduce new technology, infuse new capital, cut waste, develop new products, sub-contract inefficient operations, or make a host of other changes designed to reverse decline by increasing efficiency or innovation. Many of these adjustments require an allocation of resources to the change effort. In most cases, however, the fact of decline implies a curtailment of resources (Cameron et al., 1988), especially in the smaller, single-plant or single-function organization. When there is scarcity of available resources, managers may have little choice but to reduce the scale of operations throughout the system. Obviously, this overall scale-reduction programme usually includes a reduction in the workforce.

When considering organizational coping strategies that require workforce reductions, it helps to return to the notion of controlling the positive feedback loops that accelerate organizational decline. The key point of intervention for managers involves the linkages to and from job insecurity. Figure 8.1 (p. 163) reminds us that the precursor of job insecurity is workers' assessment of the level of organizational adaptation. In a decline situation, workers realize that the organization is performing poorly because they in fact spend a great deal of time discussing organizational effectiveness. The realization that the organization is no longer effective does not automatically imply that jobs are insecure, however. For example, governments and monopolies can be perpetually maladapted without any risk to their workers' jobs, because there is no market mechanism to force the organization to adapt to its niche. The linkage between knowledge of decline and the experience of insecurity involves expectations that managers will respond in a way that will affect the continuity of their current work situations. By shaping workers' expectations, managers can have an enormous impact on workers' experienced job insecurity, and its consequences for subsequent organizational effectiveness. Because workers' expectations are focused on the decisions managers are likely to make, it follows that the quality of decision-making will be a strong determinant of job security for the workforce.

Managers who are responsible for implementing workforce reductions must make their decisions in a context. A key contextual factor in adjusting a workforce is the labour market, the demand for and supply of workers and the processes that affect movement in and out of jobs. The most obvious conception of the labour market involves the workforce that is external to the organization. However, there is a parallel structure within the organization which can be referred to as the internal labour market. To manage a reduction of the internal workforce effectively, managers must reconcile the dynamics of these two phenomena.

Reconciling Internal and External Labour Markets

Table 9.1 shows the often overlooked fact that internal labour markets have features that are analogous to those of external labour markets. To be effective, organizational coping strategies need to take into account the dynamics of both labour markets. For example, managers cannot simply look at demand and supply conditions within the organization: they must also think about how displaced workers will fare in the external labour market, whether replacements will be available if there subsequently is an upsurge in demand, and whether a reputation for unstable employment will make it harder to recruit needed workers in the future. This figure builds on earlier conceptual work by Greenhalgh et al. (1988).

Demand and Supply
When the external demand for skills is low relative to supply, managers will tend to choose to reduce the number of workers in an

Table 9.1   *Summary of internal and external factors affecting choice of workforce-reduction strategy*

Characteristics of context	Internal labour market	External labour market
Demand and supply	Supply–demand relationship	Demand for skills
	Internal transfer opportunities	Availability of replacement skills
Controlling ideology	Organizational values	Public policy
Response routines	Standard operating procedures	Union contract provisions
	Experience with shrinkage	Restrictive legislation
Buffers	Overtime	Sub-contracting
	Cross-training	Temporary labour
	Work-sharing	
Resource supports	Slack liquid resources	Subsidies
		Protectionist legislation

oversupplied job category through internal redeployment. Natural attrition programmes do not work well in low-demand external labour markets because the 'pull' of external opportunities is minimal. For the same reasons, outplacement is difficult to accomplish and job losses put displaced workers in a difficult predicament. Experience in Europe has shown, however, that if the severance pay is lucrative enough, workers will voluntarily abandon their jobs even in high-unemployment labour markets. Thus, internal redeployment strategies such as transfer – with retraining if necessary – are the most desirable options because they do not require the costly incentives needed to induce attrition.

It should be obvious that managers in large, complex multinationals will have more options for internal redeployment. In the more common case of the single-plant, single-function organization, transfer to under-supplied job categories will perhaps take care of a few workers, retraining will not be a realistic option, and the surplus resources will be inadequate to facilitate internal redeployment.

When the external demand for skills is high, managers tend to avoid workforce-conservation strategies. Displaced workers can more easily find replacement employment, therefore managers have less of a moral responsibility to protect workers who lose jobs as a result of their decisions. Indeed, when external demand is very strong, workers can emerge from the displacement better off economically: they may move from the declining organization in which there is low job security and limited career mobility potential to a position of greater opportunity in a more effective organization.

Managers also must take into account the absolute level of supply in the external labour market, projected into the future. They need to consider this as they do their forecasting of future conditions within the internal labour market (discussed more fully below), and these considerations may favour releasing rather than retaining surplus workers when there is high demand in the external labour market. Specifically, if forecasts indicate an eventual recovery in internal demand, managers need to plan for the subsequent reacquisition of skills that are deemed to be surplus during the shrinkage programme. When the supply of requisite skills is forecast to be readily available in the external labour market, workforce displacement tends to be chosen over workforce-conservation strategies, because conservation usually has higher short-term costs.

Evidence of this decision pattern can be seen in the differential retention of highly skilled workers. In effect, managers seem more willing to take their chances on being able to attract unskilled workers when internal demand recovers than they are in the case of hard-to-replace technical or managerial personnel. As a result, workforce reductions commonly affect unskilled workers to the greatest extent, resulting in an increase in administrative intensity, the ratio of the administrative (those who support the production process) and productive (those who actually do the work) components of the workforce (Ford, 1980; Freeman and Hannan, 1975; Tsouderos, 1955).

Managers must also consider the fluidity of the external labour market. When a labour market is rigid, there is little mobility across organizations, access to employment is constrained, and attractive jobs are filled from within. The Japanese labour market is often characterized as rigid, with the culture strongly supporting lifetime employment for males in large organizations and the labour-market infrastructure largely lacking mobility-enhancing institutions such as managerial search firms and outplacement assistance specialists. On the other hand the labour market in the USA is generally fluid, with the legal structure emphasizing employment at will, prevalent norms supporting inter-organizational mobility, and an active infrastructure to facilitate 'job-hopping'. This is not to imply that all US employment is fluid. In many job categories, inter-company mobility is proscribed; access to jobs is constrained in certain union-controlled occupations such as crafts; and when an industry or region is in decline, the workforce oversupply may rigidify the labour market. In Europe, managerial discretion is constrained by a variety of employment laws and union activity which create some rigidities in the labour market. The broad picture that emerges from these considerations is that managers tend to prefer workforce-conservation strategies in rigid labour markets, and outplacement when the labour market is more fluid.

The internal labour market has its own demand and supply

characteristics. In a shrinkage situation, this means that for each job category, managers must assess current and future needs and then determine the suitability of the existing workforce to meet those needs. On the demand side, relevant issues are how great is the shortfall in demand (magnitude), how long will the shortfall last (longevity), and how certain are managers about their demand forecasts (predictability). If the firm is multidivisional, assessment of the internal demand includes investigation of opportunities to transfer surplus workers to higher-demand divisions.

On the internal supply side, issues relevant to the retention strategy – which attempts to preserve job security – are the extent to which surplus workers have generalizable versus organization-specific skills, and whether the workers are retrainable. Generalizable skills are those that can be used elsewhere in the organization and in other organizations: word-processing is a good example. Organization-specific skills are those that are unique to the functioning of the focal organization and that new entrants need to be trained or socialized to acquire. 'Learning the ropes' is a colloquial term used to refer to these skills: the newcomer may need to learn about the culture and standard operating procedures, and who are the gatekeepers, power elites and coalitions that control key organizational decisions (Van Maanen, 1976). The decision to retain and retrain involves a cost–benefit trade-off that takes into account the ratio of generic to specific skills: it is the latter that need replacing through retraining.

Forecasting is also necessary on the supply side of internal labour market analysis. Specifically, managers need to know attrition rates by job title, because there can be tremendous variability across jobs. In large organizations with few job titles, the attrition data can be very reliable, and sophisticated analysis techniques will be useful in workforce planning during decline. One such technique is transition probability matrix analysis (see Milkovich et al., 1976), which maps flows of workers into, within and out of the organization's internal labour market. Knowledge of flow rates out of an oversupplied job category is necessary in planning an attrition programme, yet few US organizations calculate simple attrition rates; fewer still can identify complex flow patterns. This ignorance in many cases leads to unnecessary job loss because the same result could be accomplished by simply not hiring replacements when people voluntarily leave.

One consequence of the internal labour market perspective is that when the shortfall in demand is of high magnitude, managers (particularly in US organizations) will tend to prefer to reduce a workforce through job loss. This preference arises because a large number of surplus workers will overwhelm attrition and redeployment strategies. In most companies, managers facing plummeting internal demand will perceive that they will have to endure the repercussions of involuntary strategies anyway, and will therefore tend to prefer the most expedient of these – job loss. Despite this preference, however, managers may be constrained from actually terminating workers, even when demand falls far short of supply. In many European countries, legislation restricts job loss, managers are required to consult trade unions and/or works councils, and surplus employees must be given substantial advance notice before they can be terminated. These factors combine to make job loss a less feasible – or at least less flexible – option outside the United States.

When the shortfall in demand is forecast to last only a short time, managers tend to choose redeployment. Redeployment within the organization (for example, transfer to other jobs, work-sharing, shortened hours or work-weeks, and assignment to part-time status) conserves the workforce until demand returns to the pre-shrinkage level.

When the shortfall in demand is unpredictable in magnitude or duration, managers will prefer to make high-magnitude cuts. Deep cuts tend to be preferred because the trauma of the more aggressive workforce-reduction tactics is increased when the process continues in steps over time. The most traumatic tactic is waves of terminations: the job-security crisis lasts a long time because even workers whose continued employment is objectively secure experience a tension they characterize as 'waiting for the other shoe to fall'. To avoid this stress, managers tend to prefer to make cuts that exceed the expected size of the shortfall. Unfortunately, the more human, single cuts are not always feasible. In the UK, for example, cuts involving more than 100 workers must be preceded by ninety-days advance notice. This requirement encourages managers to make waves of cuts of less than 100 workers in order to avoid the constraints on managerial discretion and flexibility.

The Controlling Ideology
Managers respond to the pushes and pulls of demand and supply characteristics of the internal and external labour market, but their decisions do not simply involve optimization in the face of economic dilemmas. Rather, these decisions are guided and in some ways constrained by considerations that are more ideological than economic. In this sense, economic decisions are those that simply maximize utilities, while ideological decisions are those that are guided by a value system other than profit/loss optimization, such as humanistic concerns.

The ideological emphasis of the external labour market is embodied in public policy. The high variety – across countries and regions – of laws concerning jobs at risk reflects vast difference in ideology, ranging from virtually unconstrained managerial discretion in the most extreme free-market societies (for example, in the USA) to elaborate guidelines for adjusting a workforce in societies with strong social-democratic or socialist political traditions such as, until recently, Israel and the Netherlands. Laws are only a codification of the underlying ideology; the ideology itself resides in managers' ways of thinking about their decisions, and in workers' expectations. Thus, controlling ideology goes beyond the constraining effects of what is written and enforced and extends to influence over the way the demand–supply mismatch is framed as a problem.

Public legislation has its organization-level analogue in the dominant values that shape managers' choices. For example, IBM Corporation is well known for having an organizational ideology that emphasizes conservation of the internal labour market when there is need for major organizational change.

The IBM corporate ideology arises from the basic tenet of respect for the individual. This principle, shared and understood at all levels of management, gives rise to the operating practice of full employment. Specifically, workforce adjustments must be made in such a way that no worker with satisfactory performance shall involuntarily lose employment at IBM (Greenhalgh et al., 1986). Note that as an ideological position, it falls short of being a corporate policy. Managers do not terminate surplus workers because they are not allowed to do so; they avoid terminations because they share IBM's commitment to continuity of employment.

An illustration of the opposite extreme is an account of Atari Corporation's style of achieving workforce adjustments during its experience of turmoil in the video-game industry. Prior to experiencing organizational decline in late 1982, the Atari Corporation had been wildly successful; it controlled 75 percent of the booming video-game market. But a sharp drop in demand for the firm's products was not anticipated by Atari's management. Indeed, it took them several months to notice that a drastic drop in demand had occurred. Once they realized that Atari had too many workers, they used a series of steps to reduce the workforce that reflected little concern for their workers as human beings (Sutton et al., 1986). They used only job loss to reduce the size of their workforce, and made no attempt at using natural attrition or other strategies short of job loss.

Workers were terminated in such a way that made it difficult for them to leave with dignity. Indeed, a class-action suit filed against Atari by workers asserted that the job-loss process was humiliating:

> On February 22, 1983, after the production workers arrived at work, they were told by [a production manager] that they were being laid off as of that moment. The production workers had no prior knowledge that the layoffs were going to take place. In fact, like the quality assurance workers, they had been told repeatedly that their jobs at Atari were secure. [The manager] gave the employees directions to Sunnyvale High School and told them to

be there on Friday, February 25 to pick up their final checks. He then collected their badges and the production workers were escorted off the premises. (Sutton et al., 1986: 21)

Furthermore, the lack of concern about workers as human beings also was conveyed by the absence of outplacement assistance. Atari provided severance pay for displaced workers. But only feeble efforts were made to help displaced workers find new jobs. Finally, the lack of concern about displaced workers – and the values held by Atari's leaders – was also reflected in management's negative attitudes about those who had lost jobs. The first major job loss occurred in February 1983; about 600 workers were terminated in one day. Rather than allowing workers to express sadness about co-workers who lost jobs, top management was reported to have belittled those who were let go:

> Top management went around and spoke to everybody. What they said was, 'Now we've gotten rid of all rummies, and the company's strong and all the good people are left.' And they never should have said that. They should have said, 'Because of business problems, we've had to let people go.' But they said, 'We've gotten rid of all the scum,' and that wasn't the case at all. And everybody knew it and everybody resented it. So it just got worse and worse. (Sutton et al., 1986: 23)

In short, Atari's management had an espoused policy that emphasized 'the importance of our human resources'. At the first sign of trouble, however, they demonstrated that they in practice had little concern for their workers. The use of job loss is sometimes unavoidable when the onset of decline is sudden. In this case, however, the use of humiliating means to implement terminations, the lack of outplacement assistance and the denigration of those who lost their jobs exposed Atari management's real value system. One would expect long-term consequences of this treatment of workers. It would tend to create a low-involvement corporate culture, and make it difficult to attract future workers if there were an upturn in business.

A stark contrast is the case study of the loss of jobs in 1985 at Apple Computers (Clock et al., 1986). That study shows that in very similar circumstances, including the same local culture, workers can be terminated in ways that are consistent with humanistic values. Apple let go 20 percent of its workforce during a one-week period, but workers who lost jobs were allowed to participate in parting ceremonies that let them cope with their separation from co-workers. In addition, 'surviving' co-workers were encouraged to mourn for the loss of their colleagues. Apple also spent a great deal of money on outplacement assistance. In fact, they opened an outplacement support centre that included professional counselling services for those who had lost their jobs. Furthermore, Apple leaders emphasized that the people who had been terminated were competent, valued workers and that they were sorry that Apple had to lose such fine people.

The interesting point here is that IBM, Atari and Apple were subject to the same, permissive public policy considerations. The differences among three managerial reactions seem primarily attributable to differences in organizational values. These value systems, in turn, have enormous consequences for job insecurity.

Of course, organizational values do not always determine choices. There are several case situations in which even sincere humanistic organizational values have broken down under the pressure of a severe workforce oversupply (Kodak, Polaroid and ICI are notable examples), as well as instances in which organizations have adopted humanistic values after a period of harsh economic expediency (see McKersie et al., 1981).

Response Routines
Response routines are also important determinants of organizational coping strategies. Whereas supply and demand dynamics dictate what has to be done and what general options exist, and ideology influences how humanistic the organization's workforce-reduction programme will be, response routines are the specific actions managers can take.

Response routines may be influenced from outside the organization by means of union–management contracts or restrictive legislation. A familiar example of an externally imposed response routine is the requirement of advance notice of a plant closing or other major workforce upheaval. The effect of an advance-notice requirement is to build the need for planning into the managerial decision-process. Surprisingly, planning for major organizational upheavals is often lacking in the absence of advance-notice provisions, and there have been many examples of unconscionable human-resource mismanagement practices even in otherwise sophisticated organizations. The all-too-typical scenario is that managers stumble into an oversupply situation with no forethought and, in the face of the crisis that their lack of planning has created, impose simple, heartless and wasteful solutions that are expedient in the short run and harmful in the longer run.

Other external requirements might include such things as the provision of severance pay; involvement of unions, works councils or other employee representatives in the planning process; internal redeployment activity such as inter-plant transfer; or outplacement (external redeployment) efforts. All of these external labour market requirements affect how managers reduce a workforce – either directly, by prescribing or proscribing specific tactics, or indirectly, by raising the cost of various options.

The internal labour market analogue of restrictive legislation and contract provisions is the organization's standard operating procedures (Allison, 1971). Managers and the individuals who execute their directives do not always respond in a unique way that is tailored to the exact features of each new situation. Instead, they tend to follow pre-set procedures, and these can be more or less formalized. The most formal procedures are set forth in a policy manual; the least formal are habitual ways of doing things, or simply repeating a response that was previously used in a similar situation.

The success of IBM in coping with workforce upheavals at its US plants illustrates the importance of contingency planning and the development of standard operating procedures *before* there is a crisis. IBM has formalized the procedures to be followed by managers when the workforce needs to be adjusted in response to a downturn. In addition, IBM provides human-resource specialists to help line managers implement the procedures. The result is the continuing evolution of response routines as the specialists gain experience in new situations (Greenhalgh et al., 1986).

Once again, the case of Atari provides a sharp contrast to IBM. Perhaps one reason that Atari used only job loss instead of less severe forms of workforce reduction, and perhaps one reason that little outplacement assistance was provided to displaced workers, was that the firm had no formal or informal standard operating procedures in place for responding to workforce oversupply. Thus, leaders may not have thought about using less severe workforce-reduction strategies such as reduced work-weeks or more humanistic implementation strategies such as outplacement assistance because these procedures simply were not among Atari's existing response routines.

Buffers

The view of an organization as an open system (for example, Katz and Kahn, 1978) provides insights into how it protects itself against shocks that could disturb its steady-state equilibrium. The generic protective mechanism is buffering, a process of absorbing shocks so that they cannot disrupt the technical core of the organization (see Thompson, 1967). When the focus is on the core workforce as an open system, buffering mechanisms can be identified both inside and outside the organization.

The buffers in the internal labour market are organizational coping mechanisms that avoid putting jobs of core workforce members at risk. Overtime allows the organization to increase production during peak demand periods without expanding the workforce to a size that would become a burden when production requirements subside; another tactic to avoid hiring extra people is to provide cross-training so that existing workers can perform tasks beyond their primary function. During periods of low demand for production, employees can work for fewer hours (for example, three-day weeks) so that scarce work is shared and jobs of work workforce members are not put at risk. This option may pose problems of remuneration, but avoids precipitating a job-insecurity crisis.

The buffers in the external labour market are sub-contracting and temporary labour, and these mechanisms are closely related (Pfeffer and Baron, 1988). In both cases, a peak workload is handled by workers who are not members of the organization's core workforce. In the case of sub-contracting, the task is performed by employees of another organization, whereas temporary workers are drawn into the focal organization for the purpose of achieving peak production and never become part of the core workforce. This strategy allows the organization to meet the demand for its products or services without expanding the core workforce. It thereby avoids putting jobs at risk when the peak passes. Pfeffer and Baron (1988) report that the use of both of these forms of buffers is increasing in the USA and the UK, especially among organizations that are situated in rapidly changing environments (see, for example, Atkinson and Meager, 1986; Marginson et al., 1988; Pfeffer and Baron, 1988).

The use of external labour market buffers is sometimes criticized for achieving internal labour market stability at the expense of external labour market stability. In other words, organizations using these buffering techniques are in effect passing on the problems that arise from demand and supply fluctuations to other workers and other employers (Greenhalgh et al., 1986). Two points can be made in response to this criticism. First, if an organization abandoned such buffering practices and terminated workers during oversupply periods, the general labour market would be no better off. Second, in an efficient labour market, the provision of job security within the focal organization is factored into the equity balance (Adams, 1965) of the psychological contract. Economic theory assumes that the organization wishing to buffer its core workforce would offer an equitable psychological contract, presumably providing less remuneration or requiring a higher level of worker contribution (for example, more organizational commitment) in order to compensate for the provision of job security. In practice, however, it is often the core workforce that receives both higher remuneration *and* higher job security.

Resource Supports
Workforce reduction, like any organizational change, consumes resources. Money is the obvious consumed resource, and most others have a monetary cost. Expedient solutions such as simple job loss tend to cost the organization little in the short run (unless they occur in a country which requires severance payments), even though the longer-term costs of a job-security crisis among the survivors can be enormous (Greenhalgh and McKersie, 1980). By contrast, the more human psychological-conservation strategies tend to incur high short-term costs. As a result, the choice of organizational coping strategies is at the mercy of the availability of liquid resources.

The necessary resources can come from the external labour market in the form of subsidies. In many cases in the United States, the funds have been made available without providing the comprehensive programme necessary to ensure an orderly transition to re-employment. Experience has shown that resource supports without a programme are not much better than a programme for re-employment without resources. The former postpones the suffering, the latter frustrates those who are trying to help and those who need help.

Sometimes the relief takes the form of protectionist legislation rather than direct subsidy. A declining organization cannot compete effectively because it has become maladapted. The withdrawal of market support means withdrawal of resource inputs (Katz and Kahn, 1978). Such resource-withdrawal deprives the organization of the funds necessary to conduct a workforce-readjustment programme or to reverse the decline, either through innovation or through changes that will increase efficiency. If the organization can be protected from competition – through quotas, tariffs or import restraints, for example – then the organization will have a chance to regenerate the necessary liquid slack from its own operations, rather than by means of a direct subsidy.

When the necessary funds must come from within the organization, workforce-conservation programmes are at the mercy of the slack liquid resources available to them. Unfortunately, when an organization is in decline, at that point in time it is least likely to be able to generate funds for workforce redeployment programmes (Greenhalgh, 1983). The larger and more diverse the organization, the more likely it is able to find resources to allocate to such programmes. The small, single-plant, single-function organization may be so strapped for funds that workforce-conservation programmes are not a feasible option no matter how much managers want them.

Organizational slack is a surplus of resources over what is required to maintain equilibrium in an environmental niche (Cyert and March, 1963). Slack is seen as necessary in buffering the organization's technical core from environmental adversities (Thompson, 1967). It is a key variable in the case of organizational decline, since it can be both an effect and, ultimately, a cause. Yet the phenomenon of slack, much less its relationship to decline, has not been empirically investigated to an adequate degree.

Slack is a multifaceted organizational property. Whereas the various forms of slack are affected the same way by growth, they are differentially affected by decline so that a resource imbalance tends to arise. Slack liquid resources disappear with the onset of decline and are unlikely to reappear unless and until the organization recovers. Capital assets remain under-utilized throughout the decline period, with slack increasing until they are sold off. The overall effect of decline on the amount of slack in human resources depends on the organization's reaction. The basic tendency is for human-resource slack to increase as the organization shrinks its scale of operations. This can be smooth or jagged. If the organization does nothing other than cut back production, slack in human resources will grow smoothly. A series of cutbacks, as in the common case of waves of terminations, might produce a jagged, alternating pattern of having workers stretched to their limits, then in abundance. This situation can be very costly to the organization: over-taxing a minimal workforce increases the unit cost of labour due to greater overtime and lower marginal productivity due to fatigue and stress; an overabundance of workers is inefficient because the firm pays for more labour than it needs. Either condition produces further organizational maladaptation.

The effects of organizational decline on slack resources, though complicated and under-researched, are important because different degrees of slack have an impact on organizational effectiveness, as noted earlier. First, excess slack in committed resources implies ineffi-ciency (Bourgeois, 1981). The inefficiency at best decreases the chance of generating the stream of resources, particularly cash, that are needed for recovery. At worst, the inefficiency impairs the organization's ability to compete in the market, thereby re-energizing the decline cycle. Second, the limited slack in liquid resources produces a scarcity situation that increases the likelihood of divisive organizational politics (Cyert and March, 1963; Schein, 1979), which in turn erodes the social fabric of the organization and makes recovery less likely. Finally, slack in liquid and human resources is necessary for innova-tion (Miles and Randolph, 1980), which in turn is necessary to reverse decline.

When the organization is not supported in its efforts to generate liquid slack by means of protectionist legislation, it must liquidate assets that are not crucial to its core business. Managers face a dilemma in pursuing this strategy, however. The process of selling off non-central assets in order to buffer the core workforce can increase the vulnerability of the organization as a whole in the face of environmental deterioration. Conglomerate diversification buffers the organization against the adverse effects of decline in its major niche; selling off peripheral divisions generates funds but concentrates the impact of environmental shocks in the remaining organizational sub-units.

Alternative Organizational Coping Strategies

In the preceding section we explored the aspects of the external and internal labour markets that shape managers' options for dealing with the problems of a decline situation. These considerations form the context in which managers choose a basic strategic direction for organizational coping. Preventative approaches aim to preclude a job-insecurity crisis within a workforce that is intended to be kept basically intact. Ameliorative approaches are aimed at minimizing the hardships experienced by job losers. Restorative approaches are intended to repair the damage done to the commitment and morale of the survivors of a workforce reduction. Each will be discussed in turn.

Preventative Strategies
The central objective of managers choosing a preventative approach is to avoid putting jobs at risk. The best way to avoid this predicament, of course, to to avoid organizational decline – to maintain the level of adaptation by monitoring and accommodating environmental shifts, stimulating innovation and constantly bolstering efficiency. If decline *has* occurred, however, and the organization must reduce its scale of operations in accordance with what the niche will support, managers who choose a preventative strategy can reduce the workforce in a way that minimizes job-insecurity problems.

Preventative approaches being with workforce planning. This means forecasting the trends and cycles of the organization's labour demand, developing contingency plans, and creating buffers. It also means knowing attrition rates in each job category. All too frequently organizations fail to do adequate workforce planning, even when their planning for other resources is sophisticated. Instead, they treat management of the workforce as a crude inventory problem: if they need more workers, they hire them; if they have too many, they lay some off. In some cases, hiring and termination decisions are made by different managers, with the result that hiring may be going on even as workers in the same job category are being terminated. This state of affairs would be comical were it not for the disruption of people's lives that results. Workers tend to be aware of poor workforce planning, and it aggravates their sense of job insecurity.

Figure 9.1 shows the relationship over time between workforce size and short-term payroll costs. This relationship needs to be taken into account when planning to reduce a workforce. For illustrative purposes, the diagram focuses on only one job category; the analysis needs to be repeated for each job category at risk, because attrition rates tend to vary widely across job categories, and the variability produces different slope lines. Also, the figure is based on the simplifying assumption of a sudden realization that the workforce must be shrunk from a previous steady state to a subsequent steady state. We can consider managers' two extreme options: termination and natural attrition.

The termination option means an immediate reduction in workforce size from the previous steady-state level up to the time of the decision (point D) to the desired steady-state level (point L and beyond). The

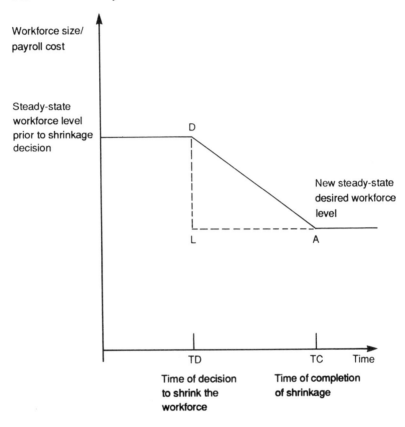

Figure 9.1 *Effects of natural attrition versus instant dismissal as a means of shrinking a workforce*

natural attrition option means that a hiring freeze is imposed at point D, so that workers who resign, retire or become incapacitated are not replaced. The workforce size atrophies to the desired level at point A, after which the new steady-state level is maintained by lifting the hiring freeze and replacing any workers who subsequently leave. The area within the triangle DAL represents the holding costs of a natural attrition programme, equivalent to the net costs as compared with instantaneous termination. The term 'holding costs' is chosen because it describes the costs of retaining workers on the payroll who are no longer needed after the decision point D.

An obvious question that arises from this perspective is, why would any economically rational manager choose natural attrition when an immediate job loss avoids the holding costs? The answer is that the analysis thus far has only taken into account short-run payroll costs. The termination option incurs longer-run identifiable economic costs

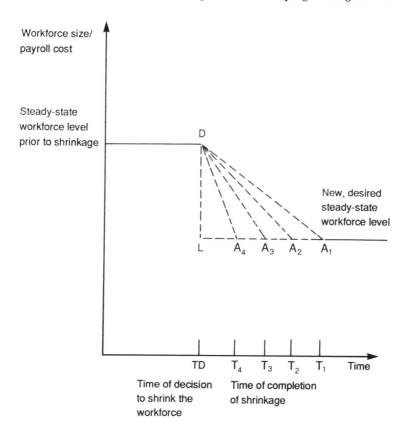

Figure 9.2 *Effects of induced attrition as compared with natural attrition and instant dismissal as a means of shrinking the workforce*

that can dwarf holding costs (Greenhalgh and McKersie, 1980). These costs arise from the adverse effects on the survivors of a termination programme, who respond in a variety of ways described in Chapter 8 and elsewhere, particularly through lowered productivity, resistance to change, loss of the most valued survivors (workers whom management intended to retain), and subsequent difficulty in recruiting high-quality workers. In unionized organizations, the adverse effects may include an increased rate of grievances, and some concerted opposition.

Consideration of the larger total costs of job loss might raise the opposite question, of why would enlightened, economically rational managers ever choose job loss? The answer involves the concept of liquid slack, discussed earlier. Many declining organizations simply do not have available the funds to pay all the holding costs, even though their managers realize that paying holding costs is a good investment when the organization is expected to survive. It should be noted that if organizational death is inevitable, long-run economic costs are irrelevant, but managers may nevertheless face some ethical imperatives to minimize harm done to the displaced workers.

The trade-off between short-run and long-run costs can be resolved by choosing workforce-reduction strategies that fall between termination and natural attrition. The effect of these strategies can be seen in Figure 9.2. The triangle DAL can be shrunk progressively by choosing strategies for inducing workers to give up their current jobs. The impact of such inducements is illustrated by the effect of moving point A to point A_2, A_3, and so on. The choice of point A_n reflects the trade-off between avoidance of long-run costs and affordability of short-run holding costs.

IBM Corporation has developed an array of options that will tailor the attrition rate to the needs of the situation without putting jobs of the core workforce at risk (Greenhalgh et al., 1986). IBM's first option is to phase out temporary workers. Attrition is assured among this group because of their three- to six-month employment contracts have a specific end date. This phase-out can be accelerated by exercising management's option to terminate the contracts immediately.

If deeper cuts are needed, the IBM plant can limit its 'vestibule' programmes, short-term employment arrangements that give potential future employees an opportunity to experience working in the organization. Designed to attract high-talent employees, these programmes primarily involve internships for college students. This tactic is less desirable than phasing out temporary workers because IBM's vestibule programmes are important in future recruiting, and their curtailment subsequently can make it harder to attract talented individuals who are crucial to the company's technological leadership. Nevertheless, sacrificing this programme is more desirable than causing a job-insecurity crisis within the core workforce.

The next options affect the plant's core workforce directly. Either by limiting workers' transfers into the focal plant from other IBM plants or by not hiring replacements, management can institute selective controls on replacing workers who leave through normal attrition. The rate of flow into the plant subsequently can be finely adjusted by increasing or relaxing the stringency of exceptions to these controls. Skill imbalances in the plant resulting from differential attrition can be corrected by retraining workers in overstaffed job functions to fill understaffed positions.

All the above procedures limit workforce inflows; the more aggressive procedures increase outflows. The first such technique would be to informally encourage surplus workers to transfer voluntarily to other IBM plants. It could be supplemented by encouraging those on leave from the plant (educational leave, for example) to resume their IBM employment at another plant. If greater outflows were needed, management could use an early retirement incentive – usually a bonus of two additional years' pay spread over four years.

A less desirable workforce-reduction technique would be to have surplus workers perform tasks that normally are contracted out. This work might include tasks such as maintenance of buildings and grounds, although such work is not a fruitful way to use highly skilled workers. It nevertheless is more desirable than some other alternatives, and its use would reflect the primacy of IBM's commitment to continuity of employment for its own workers. However, when necessary cuts are so deep that this option needs to be exercised, some workers experience some job insecurity. Greenhalgh and Rosenblatt (1984) define job insecurity as 'perceived powerlessness to maintain continuity in a threatened job situation'. Assigning workers to tasks that are different from their normal organizational roles certainly involves a discontinuity in the job situation, and begins to precipitate some of the undesirable consequences of job insecurity, such as increased turnover among the more valuable employees.

The next step would be to 'declare an open plant'. This option permits active recruitment by other IBM plants and worker-initiated relocation funded and arranged by the corporation. Note that the corporation is willing to pay relocation costs from a corporate-level fund, even though particular plants reap the benefits of being able to recruit a low-risk worker with a known performance history within IBM. If relocation costs were allocated to plants that receive the otherwise surplus IBM workers, the chargeback for full costs might serve as a disincentive: it might prove too costly to accept inter-plant transfers rather than hire new workers from the local area. Such a disincentive would make IBM's full-employment practice more difficult to implement, and would therefore threaten the job security of the IBM system as a whole.

The last technique would be used only as a last resort. These techniques involve job changes that preserve continuity of employment (that is, no worker is actually terminated) but do not accommodate workers' primary preferences. The first such step would be to reassign surplus workers to under-supplied job titles of similar rank and status. When such opportunities were exhausted, management could reassign surplus workers to jobs of lower rank and status. If all else failed, the remaining surplus workers could be reassigned to whatever positions were available in other plants.

The advantage of induced attrition – and one reason why IBM will go to great trouble to implement the technique – is that it avoids the perception that survivors' jobs might be at risk, and thereby avoids energizing the positive feedback loops that accelerate organizational decline. This is what makes it, in essence, a preventative approach.

The term 'induced' implies that whatever attrition occurs is voluntary, giving workers prediction and control over keeping or losing their jobs. These characteristics of workers' experiences are crucial from the standpoint that job insecurity is perceived powerlessness to maintain desired continuity in a threatened job situation (Greenhalgh and Rosenblatt, 1984). Workers do not experience a high degree of powerlessness if they perceive that, ultimately, it is their choice whether to take advantage of the imperfect opportunity within IBM or take their chances in the external labour market.

As noted earlier, all workforce adjustments create some degree of job insecurity. Workers can be anxious about 'losing their job as they presently know it', and thereby experience mild job insecurity (see Greenhalgh and Rosenblatt, 1984). IBM managers are well aware of this, and realize that the more aggressive the workforce-reduction strategy, the more job insecurity it creates. But they also realize that mild levels of insecurity are better than the extreme levels induced by the possibility of sudden involuntary job loss, and go to great lengths to avoid such harsh measures. Unlike many smaller, less diversified organizations, IBM has the resources and internal opportunities to select from an array of options. It should be noted, however, that alternatives to sudden dismissal are never even considered in many organizations. As a result, it is usually negligent thinking about workforce-reduction alternatives rather than lack of resources or opportunities that explains why the more humanistic strategies are rare in the absence of legislation that mandates them.

In addition to being advantageous to organizations, induced-attrition programmes can be advantageous to the surplus workers, which makes the technique practical as well as desirable. Indeed, Cornfield (1983) describes the case of a workforce reduction in the USA in which the inducements to leave the organization were so desirable that the inducements had to be allocated on the basis of seniority, a situation that is not unusual in the UK.

Ameliorative Strategies

Whereas preventative approaches are oriented toward the intact workforce, ameliorative approaches focus directly on job losers, and indirectly on survivors whose job security is at stake. Some economic and psychological damage is inevitable when workers lose their jobs: ameliorative approaches seek to limit that damage, both for humanitarian reasons and because the intensity of the job-insecurity trauma experienced by the survivors depends on what the survivors observe and vicariously experience. Factors within managers' discretion include the amount of notice displaced workers receive, the grouping of cutbacks (for example, a single cut versus waves of smaller cuts), re-employment assistance, continuation of employer support, and allowing for worker grieving reactions in response to the loss of their jobs.

Managers differ on how drawn out the termination process should be. At one extreme, the worker is terminated without any warning, and immediately escorted from the premises by a security officer. The advantage claimed of sudden termination is that it precludes sabotage or other forms of retaliation by a disgruntled worker. There is little empirical evidence, however, to support this point of view. At the opposite extreme, workers scheduled to be cut from the workforce are notified far enough in advance so that they can secure a replacement job while still employed, thereby avoiding a period of unemployment.

The often overlooked advantage of advance notice is its impact on the job security of survivors. In countries such as the USA, where until recently there has been no advance-notice requirement, the typical termination scenario involves job loss with no warning. Workers exchange stories of such events, many of them true accounts of what happened to other workers in the community. The idea that one could be surprised by instantaneous job loss is a source of worry, and contributes to job insecurity. The requirement of advance notice takes away the anxiety about the possibility of a nasty surprise. In effect, workers can tell themselves that they need not worry about job loss now; they will have plenty of time to think about it and plan for it once they know for sure. This tends to minimize ruminating about the possibility at the expense of organizational productivity.

On the other hand, if workers know that termination is scheduled in the distant future, they can concentrate their efforts on reversing the decision rather than securing alternative employment. Sometimes, this proves to be a wise strategy, especially if there is little workforce planning and the decision is haphazard. Also, if managers are responding to turbulence rather than decline (see Cameron et al., 1988), labour demand may recover during the termination notice period. Tales of such reversals circulate in retrenching organizations, and some workers scheduled for job loss will use this information as a rationale for doing nothing to help themselves.

If termination of a subset of the workforce is absolutely necessary, managers can ameliorate the negative impact by terminating all the affected workers at the same time, and then assuring the survivors that no further cuts will be forthcoming. Of course, this requires adequate workforce planning. If managers are making human-resource decisions on the basis of a crude physical inventory model, as is typical in US organizations, they are likely to find themselves conducting waves of terminations, which minimize job security. As a result, the single-cut strategy is ameliorative to job losers in that it minimizes the trauma of protracted anxiety. It is the only way to cut a workforce if the job insecurity of survivors is to be kept within manageable bounds.

Managers can ameliorate the impact of job loss by providing economic as well as job-search assistance. Economic assistance has obvious benefit to job losers, in that it minimizes their maximum economic loss. Less obvious is its benefit to the organization. Key audiences observing the company's actions are survivors and future hires. Job-insecurity trauma among these groups is a function of the perceived magnitude of the potential loss. Loss of the current job is bad enough, but total economic loss is worse. Therefore there is a long-term pay-off to what might at first appear to be a short-term economic palliative in the form of severance pay, continued health insurance or no loss of pension benefits.

Managers' efforts to ameliorate the impact of job loss is complicated by the grief reaction experienced by workers designated to be job losers (Greenhalgh, 1979; Harris and Sutton, 1986). Understanding the grief reaction sheds light on several of the difficulties typically faced by managers trying to make ameliorative approaches work. The job is a major object for many Western workers, and its loss is grieved in much the same way as loss of any other object of equivalent importance (for example, death of a close relative, loss of a home in a house-fire, and so on). The grief reaction is generic and typically undergoes identifiable stages: denial, anger, negotiation and acceptance (Kubler-Ross, 1969).

The progression through the stages is predictable for all job losers, since the grief reaction is a deeply rooted psychological process. The timing of the progression, however, is not; it varies widely across individuals. As a result, managers may find themselves needing to deal differently with different members of the same cohort of job losers. For this reason, amelioration efforts ideally would be tailored to the individual; practically, however, programmes may have to be offered that accommodate the median case.

In summary, ameliorative programmes are not as beneficial in maintaining organizational effectiveness as are preventative approaches. In addition they can be difficult to implement because displaced employees won't help *themselves* to secure new employment unless they are in the acceptance phase of the grieving process. Ameliorative programmes may nevertheless be valuable because of the message they send to survivors – that survivors need not fear total economic loss with no assistance from the organization. This assurance helps to head off a job insecurity crisis.

Restorative Strategies

Survivors of workforce reduction are the central focus of restorative approaches. These tactics are salient once the workforce reduction is complete, and managers can turn their attention to how the surviving workforce is reacting to what has transpired.

The primary issue to be addressed is job insecurity. Survivors will be wondering whether their jobs are at risk, and managers will have to communicate with the workforce to reassure them. If no further cutbacks are needed, the best way of improving job security is a guarantee that no workers will be terminated as a means of reducing the workforce. The best way is not always feasible, however, and it is worth noting that not even IBM gives workers an absolute no-termination guarantee. Rather, they assure workers that the company is committed to the practice of full employment, and will do almost anything short of corporate self-destruction to maintain the job security of its permanent workforce.

Managers need to be honest in their assurances. False reassurances may alleviate negative consequences of job insecurity in the short run, but managers will find themselves unable to gain trust and credibility in the longer run. Thus, deceiving workers about the extent to which their jobs are at risk is impractical as well as unethical.

It is worth pausing to consider how credible an assurance of future job security can be. The least believable (and perhaps most common in the USA) is when managers urge workers not to worry about further cuts because none are planned. The absence of planned future cutbacks means nothing if inadequate planning led to the previous job-loss situation. Furthermore, workers often realize that the managers who deliver such assurances often are personnel managers who do not participate in the strategic planning that will determine future workforce requirements.

More believable is a managerial commitment in which key decision-makers pledge a course of action. The pledge may be absolute or contingent, and it may be encompassing or selective. An absolute pledge is probably less believable than a contingent pledge. No manager can forecast with certainty that an organization will be in a position to provide job security beyond the foreseeable future: the ultimate uncertainty is implicit in the definition of 'foreseeable'. Therefore a contingent pledge will be more realistic, and more convincing. The contingency may involve affordability, organizational survival or applicable legislation; the pledge may be a statement of managerial priorities, such as of worker interest over stockholder interests, of the rights of surplus current workers over potential new hires, or of job security over profitability.

An encompassing job-security pledge, whatever its contingent nature, applies to the workforce as a whole. A selective pledge applies to a subset. The subset may be permanent workers, workers on the payroll as of a certain date, or workers in a particular job category, plant or geographic location. A system of job rights based on seniority is a flexible selective pledge. The pledge is strongest to those with the most years of service, and decreases thereafter.

Seniority-based job-security provisions have disadvantages as well as advantages that make them so prevalent. A major disadvantage is that they can counteract equal-opportunity objectives. Because equal-opportunity policies were only recently implemented, workers hired under this programme – primarily ethnic minorities, women and the handicapped – tend to have low seniority compared to the workforce as a whole. As a result, they tend to be the first to be terminated in a seniority-based workforce reduction (see, for example, US Commission on Civil Rights, 1977).

Another serious disadvantage is that a seniority system is blind to merit, potential and human capital. The lowest-seniority workers must be terminated no matter how valuable they are to the organization. This may cause the loss of workers who are the most youthful and energetic, the most recently educated and therefore perhaps best exposed to the cutting edge of relevant knowledge, the most mouldable, and sometimes those in whom the organization has invested the most in special training. This latter point runs counter to classical human capital theory (for instance, Parsons, 1972; Feldstein, 1976), which assumes a monotonic relationship between seniority and accumulated human capital. Perhaps the apparent contradiction is explainable in terms of the impact of high technology on modern organizations, in contrast to the craft system of earlier times.

Given these disadvantages, it is hard to imagine why any organization would choose a seniority system to reduce feelings of job insecurity rather than a merit system. However, there are several problems with merit systems which can make them unpopular with unions, workers and even managers.

First, merit is hard to assess with a degree of certainty and demonstrability that would survive an adversarial proceeding, and workers whose jobs are at risk will vigorously contest a managerial evaluation that results in their termination. Second, even if performance appraisal is objectively accurate, it may not be subjectively credible to affected workers. As noted earlier, cognitive dissonance arises when workers see themselves as worthy of retention, but also see managers viewing them as items the organization needs to get rid of. Third, a seniority system involves publicly available easily verifiable data, in contrast to the confidential data involved in a merit system. These characteristics provide two elements that reduce the stress of a workforce reduction – prediction and understanding. Workers can assess the degree of risk to their jobs simply by knowing the depth of the workforce reduction and their place in the seniority rankings. The procedure also is easily understandable. There are no multifaceted managerial choices to understand that accommodate weightings of performance, potential and human capital. Rather, the last person hired is always the next one to be let go.

The pledge assuring job security is in essence a personal commitment and may not continue beyond the term in power of the individual(s) making the pledge. The commitment of an individual (or power elite) is not binding on successors. This is true even when commitments are publicly made, unambiguously understood and confirmed in writing. Successors can impose their own values and workforce-management practices, or they can view the situation they inherit as being different from the situation faced by the preceding regime, and on that basis declare previous agreements inapplicable. Thus, a job-security pledge based simply on personal commitment has limited value as a restorative strategy because survivors will realize its vulnerability to changing times and events.

In order to transcend individual tenure in a power situation, a job-security guarantee needs to be institutionalized. Institutionalization can involve, in increasing order of effectiveness and therefore restorative value, organizational culture, contracts, operating procedures, policies and laws.

A culture shift involves norms, values, roles and myths that reinforce continuity of employment. This approach infuses the principle of job security throughout the organization, and pervades the thinking of decision-makers and the expectations of their audiences. IBM and many Japanese organizations have such cultures.

The traditional Japanese work culture influences both employer and worker. Japanese workers may experience an almost feudal loyalty to the organizations that employ them. Continuity of employment is infused into the way they conceptualize work. In the classic case, the Japanese worker completes his or her education, then begins employment in the company he or she will retire from some forty years later. Such employers, particularly large organizations, have complementary views of job candidates. They are selecting a lifetime member of the company, with all the expectations thereby entailed.

There is some evidence that this culture is undergoing some transformation. Foreign firms operating in Japan bring their native work cultures to the Japanese labour market, and multinational Japanese companies operating in foreign labour markets and assimilate workers who were socialized in these foreign cultures (Kanabayashi, 1988). However, despite the gradual changes, the Japanese work culture remains very strong in many large Japanese organizations.

Contracts can be made at the individual or collective level, and may preclude or simply raise the costs of involuntary job loss. A generous severance-pay clause in an employment or labour contract, for example, may be expensive enough to produce immediate progress in workforce planning and development of induced-attrition programmes, thereby obviating the need for a guarantee of no job loss.

Another way to institutionalize job security is to create standard operating procedures (SOPs) that lead managers to reduce a workforce in a way that avoids involuntary job loss. Allison (1971) has noted the resiliency of an organization's SOPs when managers are trying to *change* institutionalized practices. This phenomenon can be used to advantage when decision-makers create routines – and therefore work habits – that would have to be disrupted if managers were to reduce a workforce in a way that would impair job security.

The most potent means of institutionalizing future job security is by prohibiting involuntary job loss, by means either of corporate policy or of legislative edict. Both approaches can be prescriptive or proscriptive, and are backed by legitimate and coercive power arising from the organization or the social system. In such cases, a pledge of continuity of employment is irrelevant; managers are given no choice.

The success record of attempts to restore job security in the absence of institutionalization is dismal. Workers tend not to believe managers' good intentions alone, even when these are sincere. Good intentions typically are formed in the aftermath of a job-loss débâcle, but the future decision will be made under severe stress, and in the face of powerful new economic pressures. The temptation to settle for expedient short-term solutions will once again become immensely attractive, and the good intentions can easily become overwhelmed.

Managers wishing to conduct restorative programmes must keep in mind how information is actually transmitted and shared during a workforce reduction. Official communication by managers is only a small part of the information to which survivors attend and respond. Rumours, media accounts, union polemics and myths typically account for a larger share of the relevant data.

Rumours are probably inevitable no matter how much official communication there is. Some rumours, if believed, will create job insecurity. Workers respond on the basis of their subjective reality, therefore it is unwise and unproductive for managers to dismiss rumours as unimportant. A better approach is to create a management point of view proactively, and use this to correct inaccurate and destructive rumours.

Managers often overlook the role of the media in shaping perceptions of a decline situation and workforce reduction, and therefore fail to properly influence this set of messages. The local news media are understandably interested in changes in the fortunes of employers in the area: there is widespread reader interest in prominent local organizations as well as the more specific interest of workers whose livelihoods could be affected, directly or indirectly, by significant changes in the organizations that employ them.

The union's messages regarding jobs at risk may be a particular problem for managers seeking to avoid a job-security crisis (see Chapters 5, 6 and 7 for a more complete discussion of unions' responses). The union faces its own dilemma in issuing messages. On the one hand, this institution needs to emphasize the risk to jobs, so that members will value and support the union as their protector against managerial decisions that might hurt them. On the other hand, the union needs to give the impression that jobs are secure, to show that it is an effective protector of continuity of employment. The result often is confused messages, and this situation provides an opportunity for managers to insert their own messages in order to optimize job security.

The final factor to be taken into account is the creation of myths. A myth is a story, shared among the workforce, that typifies some aspect of the organization. Myths underlie organizational culture, can be supportive or destructive, and are the substance of what organizational anthropologists might describe as the oral tradition of the workplace (Sykes, 1965). They are central elements involved in the 'social negotiation of reality'.

An example of a supportive myth is the story of a fifty-five-year-old, loyal immigrant worker who was scheduled for job loss. His organizational tenure was six months short of the fifteen years of service necessary for pension vesting (a feature of US employment agreements whereby the worker, if laid off, does not have the right to 'keep' the employer's contribution to his/her pension until a specified length of service – in this case, fifteen years). Instead of laying him off on schedule, the company created a role to keep him on the payroll until he secured full rights to his accumulated pension benefits. This story was told and retold by workers when explaining the 'family culture' that existed before the company was acquired by a conglomerate.

Three points are relevant. First, the truth or falsity of a myth is not important, only the extent to which it typifies organizational predispositions as workers understand them. Second, myths can be created, nurtured or shaped by managers who are attuned to the symbolic impact of their actions. Third, the antidote to an unfavourable myth is not official denial but additional myths that embody the opposite message. Several positive myths seem to be required to counteract a negative myth concerning the organization.

Organizational Obligations

This last section considers the resources the organization should expend in helping the workforce adjust to the organizational shrinkage programme. The organization's expenditures can be motivated by self-interest, compliance with laws or contracts, or the need to discharge a moral obligation.

Economic self-interest is a consideration because the cost of a programme to help the workforce during the upheaval can be considered an investment. The pay-off takes several forms: mitigating the impact of job insecurity among survivors, shaping organizational culture in a way that promotes workforce contributions, creating a favourable impression among potential future recruits, and enhancing the organization's image among customers, regulators and the investment community (Greenhalgh and Kaestle, 1981). Unfavourable publicity is particularly damaging when it involves human-resource management practices because audiences may easily identify with the plight of affected workers. Thus, an investment in helping workers whose jobs are at risk can be a form of institutional advertising.

Beyond the pay-off motive, there is an obligation to workers that arises from the employer–employee social contract. In essence, the employer ought to provide assistance in some form in order to discharge its economic responsibility to surplus workers who permanently lose their jobs 'through no fault of their own'. It is in effect a payment made to terminate an explicit or implicit employment contract.

Whatever the motive, managers need to take some constructive action to influence the job security of their workforce when jobs are at risk. Organizational coping strategies work best if they are preventative in nature. If job loss cannot be avoided, then ameliorative strategies are needed to cushion the impact on job losers, and to manage the reactions of the surviving workforce. In the aftermath of a workforce reduction, strategies that restore job security are important determinants of future organizational effectiveness.

10

Conclusions

Jean Hartley, Dan Jacobson, Bert Klandermans
and Tinka van Vuuren

Since recent years have been almost universally characterized as an era of rapid change, one would expect job security to be a basic variable in the socio-psychological study of organizations. It is not. (Greenhalgh, 1983: 252)

The current studies have gone some way to make up for this deficit in our understanding of job security and insecurity. We have examined the individual antecedents and consequences of job insecurity, and the impact of job insecurity on organizations and industrial relations. Our research studies cover four countries, the public and private sectors, and a variety of questions about how insecurity impinges on organizational life. Here, we take stock of what we have achieved and make recommendations about what remains to be done in both research and policy terms.

Uncertainties about the future of one's job may arise in many settings where change and ambiguity exist. Although our studies are based primarily in settings of organizational decline, it is clear that it is *uncertainty* which engenders insecurity about employment. In other words, it may occur wherever change occurs, whether that is change through recession or through restructuring, in technological development or organization development. It may even exist in situations of growth as well as in situations of decline. Borg and Hartley (1989) analysed the responses of white-collar workers working in eleven growing, 'healthy', high-tech organizations and found that job insecurity existed even there and was systematically related to lower satisfaction, commitment and job involvement.

This should not surprise us. Jacobson (Chapter 2) argued cogently that event uncertainty can be very stressful in a variety of settings, whether that is the uncertainty of a partner 'missing in action' in war, or the relief expressed by some job losers when uncertainty turns to reality. The stress literature emphasizes the general relationship between stress and uncertainty (Lazarus and Folkman, 1984), between stress and change (Dooley and Catalano, 1980), and between stress and lack of perceived control (Frankenhauser, 1981; Karasek, 1979).

As change and uncertainty is now endemic in many organizations and settings, job insecurity is a major cause for concern in the future. In Chapter 1 we showed that organizations are restructuring, introducing new technologies, internationalizing their production, experiencing more rapid changes in products, markets and fashions, and providing some employees with less secure employment. Whether the changes result in decline or growth for the organization, they spell uncertainties for at least some of their employees. Given this magnitude of change, job insecurity needs to become more central in organizational research. Yet it is generally absent. We have a large literature dealing with organizational change and organizational development which is silent about the effect of uncertainty on employees other than to say that 'resistance to change' may develop. As well as being naively managerial, on its own this perspective tells us nothing about how employees perceive and respond to what can be important changes in their psychological contract with the organization, or what some of the individual and organizational consequences might be.

Job insecurity should perhaps be seen as a manifestation of more general uncertainty a person experiences throughout his or her life in a modern society. Gone are the old certainties and stabilities of family life, marriage, community, religion and work. Instead, many people face considerable change and uncertainty in their relationships (as divorce rates show), in their place of living (with higher rates of social and geographical mobility) and in their employment (with higher levels of job change and the abandonment of 'a job for life'). Examining how people perceive and cope with job insecurity may help us to understand how other life uncertainties are experienced. And we also may need to recognize that job insecurity can place further strain on an individual already coping with other uncertainties.

The Impact of Job Insecurity

The research indicates that job insecurity can be dysfunctional for both individuals and organizations. Employees have a variety of perceptions about the same economic and organizational conditions and there are also differences in their responses to these. However, the research suggests a consistent relationship at the aggregate level, between perceptions of the job being at risk and lowered satisfaction with a range of job and organizational features. Those who are more insecure are more likely to report lower job satisfaction, lower trust in management, lower organizational commitment and greater interest in leaving the organization. Insecure employees are also more likely to report psychosomatic symptoms and to feel depressed.

The impact spreads into negative consequences for the organization. Job insecurity, far from facilitating change, actually inhibits it, because fear for the future is accompanied by resistance to change. And we have seen the far-reaching consequences for industrial relations too, with a cycle of mistrust, blame, dissatisfaction and short-term solutions, with a weakening of the abilities of either management or workforce to engage in constructive change.

On almost all measures of attitudes and opinions, insecurity is negatively correlated. Also, trust in management (found in all the studies reported here) is lower for insecure workers. Borg and Hartley (1989) found that insecure workers also rated the company's products more negatively. The implications of lower satisfaction with most aspects of the organization, found in our studies, suggest that the impact on organizational image may be profound. All employees have contacts with others outside their organization, whether through their job, through their trade union or profession, or through family, friends and neighbours. Each employee is, informally, an 'organizational ambassador', contributing to impressions in the wider world about the organization. Therefore, quite apart from human-resource management considerations, organizations need to be wary of high levels of job insecurity, since damaging messages may leak out about the organization. Organizations need to recognize that their actions in a variety of fields have indirect consequences for the image of the organization. As demographic and labour-market changes create a shortage of skilled workers over the next decade, the reputation an organization has for caring for its employees (or not) may be critical in recruitment and retention.

Research Issues

There are critical questions arising from the research which are unresolved and which raise wider issues for the direction and emphasis of work and organization psychology. The research studies, for example, have documented the impact on and coping strategies of individuals experiencing job insecurity. We have not been able to determine causal relationships because of the cross-sectional nature of the research. Does job insecurity cause lower job satisfaction and well-being? Or do employees with lower job satisfaction experience more job insecurity? Then again, are both job insecurity and negative work attitudes caused by something else, such as organizational decline or major change? Finally, it is quite possible that the relationship between insecurity and work attitudes operates in both directions, creating a reinforcing cycle of insecurity and negative outlook. Further research needs to investigate these possibilities. Longitudinal research designs, following employees through an extended period of organizational uncertainty would be valuable. The key factors in such a cycle would also need to be established – when and how does uncertainty develop? What is the role of powerlessness in engendering insecurity? What actions by managers are most likely to induce insecurity?

There is also a need for further investigations of individual variations in perceived job insecurity. Surprisingly, our research suggested that demographic and occupational characteristics were only weakly associated with insecurity. While employees' concerns were, in part, related to the objective conditions of the organization, there was also a high degree of variation in insecurity. The Dutch and Israeli studies explored the impact of pessimism, locus of control and neuroticism, finding that these were related to feelings of insecurity. However, further research into the cognitive and emotional processes underlying job insecurity would be valuable.

The studies reported in this book have focused almost exclusively on male workers. There have been too few women in these studies for separate or comparative analysis. The pattern of full-time employment without interruption is predominantly a male pattern and, for example, it could be that men as a group react to and cope with job insecurity differently from women. We do not wish to suggest by this that job insecurity is less important for women. In a related field, the idea that unemployment is somehow less significant for women has been sharply challenged (Hartley and Mohr, 1989; Martin and Roberts, 1984).

How will increasing job insecurity affect our theories and models of organizational behaviour? Many current theories are premised on an implicit model of stability and are only slowly being modified to cope with a more fluid and uncertain environment. For example, many theories of selection, assessment, career development and vocational guidance have been implicitly based on the concept of the stable career. What are the implications for theory and practice of insecure and changing employment and work roles? Currently, much education, training and skill development is based on the postponement of gratification until long and arduous training has been acquired. But why invest in an education if your job will become obsolete? What's the point of sacrifices to get trained if you know your 'job expectancy' is limited? Too high a level of job insecurity could generate a lack of motivation or skill to make the major changes which will be needed if organizations are going to stay in business. So the question is how to educate people for versatility. Furthermore, theories of career choice, recruitment and selection, training, job design, motivation and reward systems, management development, employment counselling and employment discipline among others may all need fresh understanding when job insecurity is a major component of work experience.

Policy Issues

In suggesting ways in which change and insecurity can be handled constructively, we must first recognize the constraints under which firms in capitalist societies operate. Organizations have limited scope to *prevent* job insecurity, because increasing competition, especially as markets internationalize, means that innovation and profit have to be pursued if the firm is to survive. In Chapter 1 we documented many of the pressures on firms which have generated increased insecurity for employees. Public sector organizations, operating with limited financial support, face similar pressures to keep labour costs down. As firms grow or decline, labour costs are modifiable in ways which fixed assets cannot be or take longer to achieve. The pursuit of short-term profits means that organizations often have little choice but to shed labour to survive.

A key question is whether, in some cases, a longer-term perspective can be justified, whereby the dysfunctional consequences of insecurity for the organization, which we have documented, outweigh the shorter-term gains. A further important issue is whether other individuals, groups and organizations should have some influence over the firm's decisions which generate job insecurity.

The Value of Employment Security
First, perhaps, we might assess the value of job insecurity. Is it necessarily a bad thing anyway? If change and restructuring are so prevalent and since not all employees suffer from the negative effects of insecurity, perhaps it should simply be accepted as a necessary consequence of change? Indeed, it may even have some beneficial effects. For those convinced of the supremacy of the market in generating efficiency and productivity, job insecurity can be seen as a necessary motivator and a spur to better performance.

In addition, it can be argued that too much security in a given job can lead to a number of problems for both individuals and organizations, such as difficulties of creating change, problems of motivating staff, abuse of conditions of employment and inefficiencies of output.

However, we would argue that the antidote to organizational problems of tenure is not to institute job insecurity, which brings other problems in its wake. We argue for the concept of *employment security* but without the rigidities introduced when employment is taken to be the *particular job* the person is currently doing. In this view, an employee might be offered a long-term or rolling contract with the organization, but without any guarantees that the person will necessarily remain in one specific job.

With a background of employment security, and with a clear training and development policy (a point on which many organizations would need to pay more attention), employees should be more prepared to accept change, because fears of redundancy are lessened. Organizations would then be better able to deploy their workforces across a range of tasks and activities and they would avoid the costly implications (in financial, human-resource and public-relations terms) of workforce-reduction exercises. Flexibility and a reasonable degree of employment security are not incompatible.

Employees can thus become involved in change knowing that their baseline income and patterns of life are preserved and with the opportunity for further training and skill development. Of course, in such an employment contract which emphasized employment but not specific job security, the trade unions would need to be involved (where they existed in the organization), in order to ensure that the procedures for personnel development, training and job allocation were fair and reasonable. Many trade unions would welcome the opportunity to preserve employment during periods of change, where their members had opportunities for new skills and new horizons. Of course, this view of employment security also involves a different view for many managements: a view of employees as a scarce and valuable long-term resource within a changing organization rather than a short-term cost and a drag on flexibility. Although such a view of employees is becoming more prevalent (Gutchess, 1985; Rosow and Zager, 1984), it is far from being the norm yet. The argument for employment security is not suggesting that insecurity can be *eradicated*. But we suggest that it can be *reduced* through planning for employment security. Also, where it exists it can often be *managed*.

As the stress literature repeatedly emphasizes (Lazarus and Folkman, 1984), a key for those experiencing a stressful event is to gain some sense of empowerment in relation to the event. This involves a change from feeling a victim to experiencing some degree of mastery, which involves the processes of understanding, prediction and control. These give the person some information about what is happening, why and what may be done about it. Of course, it can be psychologically damaging to suggest a person can exert control over uninfluenceable events, but there are ways to assist people to maintain realistically their self-esteem and sense of agency, during periods of uncertainty. Kobasa et al. (1982), for example, have proposed the concept of hardiness to explain why some people cope better with stressful events than others. The components of hardiness are a sense of personal influence over events and outcomes, a sense of commitment or generalized sense of purpose and a sense of challenge, defined as 'the belief that change rather than stability is normal in life and that the anticipation of changes are [sic] interesting incentives to growth rather than threats to security' (Kobasa et al., 1982: 170). Although they treat hardiness as a personality disposition, there seems no reason why hardiness may not be learned or developed.

The role of management in preventing, managing and ameliorating job insecurity has been made clear in Chapter 9 and we will not go over that ground here, though we may simply note that it is a highly pertinent area and that overall management has both the responsibility and the capacity to have a significant impact on levels of job insecurity experienced.

Where there is uncertainty, especially if its onset is sudden, management may react in an atmosphere of stress, confusion and blame. Often the last thing many managements are capable of at that stage is to help employees understand, predict and control what may happen to them. It is seen as too great a risk to admit to difficulties which open up managerial power and legitimacy to question (Cressey et al., 1985) or to engage in activities which share with employees some of the questions for the future. Yet, paradoxically, the evidence (for example, Hirschorn and associates, 1983) suggests that giving employees an understanding and involvement in employment affairs which affect them may be beneficial to the organization and head off damage to trust, commitment and innovation.

Trade unions and works councils have an important role to play in promoting employment security as we noted earlier, both through workplace negotiation for investment in skills-training and through public campaigns to promote protective employment legislation and national investment in training. Where job insecurity already exists, the role of the trade union may be highly relevant in initiating discussion, sharing information and either allaying fears or providing a realistic basis for concern. The sense of powerlessness and isolation which can accompany job insecurity means that trade unions can be particularly important in reducing these negative effects. However, we also saw (Chapter 7) that unions can be limited in what they can achieve at workplace level. Handling insecurity needs to be conducted primarily by full-time officials or through a regional or national campaign since workplace union organization may not be able to promote strategic discussion of job insecurity (see also Edmonds, 1984). However, a local union could perhaps engage in opening up issues of insecurity for discussion, ensuring that members are not trapped in individual fears and a sense of futility.

Job insecurity, though, is not simply an issue for management, workers and unions. It is too important and too complex a matter to be left solely to those inside the organization. If the organization spirals into further difficulties leading to run-down, redundancy or closure, then other groups suffer the repercussions. Outside agencies therefore have an interest in constructively tackling job insecurity at an early stage. In particular, we point to a role for government, local authorities, pressure groups and educational establishments in the management of change and uncertainty.

Central government has a key role, through the legislation it adopts to facilitate change and restructuring. Some countries have legislated to provide compulsory compensation to employees for job losses, thereby to some extent encouraging employers to think carefully before embarking on job loss as a response to change. Some job insecurity will be lessened by this legislative framework. Laws to require information-disclosure and consultation with trade unions or works councils is also a step to reducing levels of job insecurity by ensuring that employees are kept informed of issues affecting them and that they are given some influence over decisions. West Germany is a country with a high rate of innovation and change and yet also a strong legislative framework to inform and involve workers and their unions in change (Streeck, 1985).

Local government has a constructive role to play in encouraging employment security. Should job loss occur, it may represent a cost-saving to the firm but those costs are often displaced on to the community. And job insecurity itself may have some wider impacts, for example on health-care provision. Local authorities therefore have a legitimate role in collecting information about economic development, promoting discussion and in planning change. The social audit provides a way of estimating the social as well as the economic costs of restructuring (Geddes, 1988). It could include estimates of the health costs of living with job insecurity. Local authorities can be important in promoting a *public* discussion of the local economy, thus encouraging organizations to open up discussion to a wider public – including their own employees. In addition, local authorities may be able to encourage more planful approaches to change (Benington and Stoker, 1989) through economic development. This access to information, public debate and constructive planning could contribute to preventing or anticipating job loss and easing some feelings of job insecurity for those still in work. It breaks through the loneliness and isolation of the job insecure by encouraging the sharing of feelings of job insecurity and in promoting the understanding, prediction and sense of control which make transition easier.

Pressure groups can also reduce feelings of insecurity by planning constructively for change. For example, in Britain, the national Campaign for Nuclear Disarmament, in its attempts to reduce armament manufacture, engaged in detailed discussions with both management and workforce of a major weapons manufacturer in order to explore alternative uses for the workforce skills. There was a recognition of the need for continuing employment but with a different focus and product. Environmental groups, political parties and others with an interest in local employment can have a role in making transition planful and with a regard for the feelings of insecurity raised for the workforce.

Educational establishments, from schools to universities, can help to promote employment security and a positive approach to planned change. We noted earlier how investment in a career, with emphasis on postponed gratification, may militate against a preparedness to change. If educational establishments were more flexible in entry qualifications, recognizing the potential of mature students, the relearning and retraining required as a corollary of employment security might be less difficult. Also, education has an important role to play in developing confidence and self-esteem in the face of change.

Overall, we have made proposals for policy initiatives which address both the structural bases of job insecurity and the psychological reactions to job insecurity. The increase in change and restructuring, with its higher levels of uncertainty and insecurity, raises many issues and challenges for organizational psychologists, personnel and senior managers, trade union representatives and employees. Insecurity is also a matter of concern to the wider community. Our main proposal is for greater consideration to be given to employment security, with training and development for employees as they move from job to job. In this way industrialized societies can meet the challenge of change and innovation while considering the needs and legitimate expectations of their workforces.

References

Abramson, L.Y., Seligman, M.E.P. and Teasdale, J.D. (1978) 'Learned helplessness in humans: critique and reformulations', *Journal of Abnormal Psychology,* 87: 49–74.

ACAS (1988) *Labour Flexibility in Britain: The 1987 ACAS Survey.* Occasional Paper No. 41. London: ACAS.

Adams, J.S. (1965) 'Inequity in social exchange', in L. Berkowitz (ed.), *Advances in Experimental Social Psychology.* New York: Academic Press. Vol. 2, pp. 267–99.

Albert, S. (1984) 'A delete-design model for successful transitions', in J.R. Kimberly and R.E. Quinn (eds), *Managing Organizational Transitions.* Homewood, Ill.: Irwin.

Allison, G.T. (1971) *Essence of Decision.* Boston: Little, Brown.

Allport, F.H. (1933) *Institutional Behaviour.* Chapel Hill: University of North Carolina Press.

Antaki, C. (1981) *The Psychology of Ordinary Explanations of Social Behaviour.* London: Academic Press.

Applebaum, E. (1985) 'Alternative work schedule of women'. Unpublished manuscript, Department of Economics, Temple University.

Argenti, J. (1976) *Corporate Collapse: The Causes and Symptoms.* Maidenhead: McGraw-Hill.

Argyris, C. (1960) *Understanding Organizational Behaviour.* Homewood, Ill.: Dorsey.

Ashford, S.J., Lee, C. and Bobko, P. (1987) 'Job insecurity: an empirical analysis of three measures'. Paper presented at the Annual Meeting of the Academy of Management, New Orleans.

Ashford, S.J., Lee, C. and Bobko, P. (1989) 'Content, causes, and consequences of job insecurity: a theory-based measure and substantive test', *Academy of Management Journal,* 32: 803–29.

Atkinson, J. and Meager, N. (1986) 'Is flexibility just a flash in the pan?', *Personnel Management,* September: 26–9.

Bacharach, S. and Lawler, E. (1980) *Power and Politics in Organizations: The Social Psychology of Conflict, Coalitions and Bargaining.* San Francisco: Jossey-Bass.

Baker, G.W. and Chapman, D. (1962) *Man and Society in Disaster.* New York: Basic Books.

Barnard, C.I. (1938) *The Functions of the Executive.* Cambridge, Mass.: Harvard University Press.

Bassett, P. (1986) *Strike Free: New Industrial Relations in Britain.* London: Macmillan.

Batstone, E. (1984) *Working Order.* Oxford: Blackwell.

Batstone, E., Boraston, I. and Frenkel, S. (1977) *Shop Stewards in Action.* Oxford: Blackwell.

Batstone, E., Boraston, I. and Frenkel, S. (1978) *The Social Organization of Strikes.* Oxford: Blackwell.

Batstone, E. and Gourlay, S. (1986) *Unions, Unemployment and Innovation.* Oxford: Blackwell.

Becker, H.S. (1960) 'Notes on the concept of commitment', *American Sociological Review,* 66: 32–40.

Benington, J. and Stoker, G. (1989) 'Local government in the firing line', in N. Buchan (ed.), *Glasnost in Britain?* London: Macmillan.

Berren, M.R., Beigel, A. and Ghertner, S. (1986) 'A typology for the classification of disasters', in R.H. Moos (ed.), *Coping with Life Crises.* New York: Plenum Press.

Beynon, H. (1973) *Working for Ford.* London: Allen Lane.

Blackaby, F. (1979) *Deindustrialization.* London: Heinemann.

Blake, R. and Mouton, J. (1962) 'The intergroup dynamics of win–lose conflict and problem-solving collaboration in union–management relations', in M. Sherif (ed.), *Intergroup Relations and Leadership.* New York: Wiley.

Block, R.N. (1977) 'The impact of union-negotiated employment security provisions on manufacturing quit rate', in *Proceedings of the Twenty-Ninth Annual Winter Meeting,* Industrial Relations Research Association, Madison.

Bluestone, B., Harrison, B. and Clayton-Mathews, A. (1986) 'Structure vs. cycle in U.S. manufacturing job growth', *Industrial Relations,* 25: 101–17.

Borg, I. and Hartley, J. (1989) 'Job insecurity and work attitudes'. Unpublished paper, Occupational Psychology Department, Birkbeck College, University of London.

Bourgeois, L.J., 3rd (1981) 'On the measurement of organization slack', *Academy of Management Review,* 6: 29–39.

Brett, J.M. (1984) 'Job transitions and personal and role development', *Research in Personnel and Human Resources Management,* 2: 155–85.

Brief, A.P. and J.M. Atieh (1987) 'Studying job stress: are we making mountains out of molehills?', *Journal of Occupational Behaviour,* 8: 115–26.

Brief, A.P. and Oliver, R.I. (1976) 'Male–female differences in work attitudes among retail sales managers', *Journal of Applied Psychology,* 61: 526–8.

Brief, A.P., Oliver, R.I. and Aldag, R.J. (1977) 'Sex differences in preferences for job attributes revisited', *Journal of Applied Psychology,* 62: 645–6.

Bright, D., Sawbridge, D. and Rees, B. (1983) 'Industrial relations of recession', *Industrial Relations Journal,* 14: 24–33.

Brim, O.G. (1980) 'Types of life events', *Journal of Social Issues,* 36: 148–57.

Brockner, J., Davy, J. and Carter, C. (1985) 'Layoffs, self-esteem, and survivor guilt: motivational, affective and attitudinal consequences', *Organizational Behaviour and Human Decision Processes,* 36: 229–44.

Brockner, J., Greenberg, J., Brockner, A., Bortz, J. Davy, J. and Carter, C. (1986) 'Layoffs, equity theory, and work performance: further evidence of the impact of survivor guilt', *Academy of Management Journal,* 29: 373–84.

Brockner, J., Grover, S., Reed, T., DeWitt, R. and O'Malley, M. (1987) 'The effect on survivors of their identification with, and the organization's compensation to, the victims of layoffs: converging evidence from the laboratory and field', working paper, Graduate School of Business, Columbia University.

Brown, R. (1988) *Group Processes: Dynamics within and between Groups.* Oxford: Blackwell.

Brown, W. (1973) *Piecework Bargaining.* London: Heinemann.

Brown, W. (1983a) 'The impact of high unemployment on bargaining structure', *Journal of Industrial Relations,* 25: 132–9.

Brown, W. (1983b) 'Britain's unions: new pressures and shifting loyalties', *Personnel Management,* October: 48–51.

Brown, W. (1986) 'The changing role of trade unions in the management of labour', *British Journal of Industrial Relations,* 24: 162–8.

Brown, W. (1987) 'Pay determination: British workplace industrial relations 1980–84', *British Journal of Industrial Relations,* 25: 291–4.

Brown-Johnson, N. (1987) 'Conceptualizing unions' influence on members' perceptions of job insecurity'. Unpublished paper, Department of Management, University of Kentucky.

Buchanan, B. (1974) 'Building organizational commitment: the socialization of managers in work organizations', *Administrative Science Quarterly,* 19: 533–46.

Burke, R.J. (1966) 'Are Herzberg's motivators and hygienes unidimensional?', *Journal of Applied Psychology,* 50: 317–21.

Burr, W.R. (1972) 'Role transition: a reformulation of theory', *Journal of Marriage and the Family*, 34: 407–16.

Büssing, A. and Jochum, I. (1986) 'Arbeitsplatzunsicherheit, Belastungserleben und Kontrollwahrnehmung; Ergebnisse einer quasi-experimentellen Untersuchung in der Stahlindustrie', *Psychologie und Praxis: Zeitschrift für Arbeits- und Organizationspsychologie*, 30: 180–91.

Buunk, A.P. and de Wolff, C.J. (1989) 'Sociaal psychologische aspecten van stress op het werk', in P.J.D. Drenth, H. Thierry, P.J. Willems and C.J. de Wolff (eds), *Nieuw Handboek Arbeids- en Organisatiepsychologie*. Deventer: Van Loghum Slaterus.

Cameron, K., Sutton, R. and Whetten, D. (1988) 'Issues in organizational decline', in K. Cameron, R. Sutton and D. Whetten (eds), *Organizational Decline: Frameworks, Research and Prescriptions*. Boston: Ballinger Publishing.

Cameron, K.S., Kim, M.U. and Whetten, D.A. (1987) 'Organizational effects of decline and turbulence', *Administrative Science Quarterly*, 32: 222–40.

Caplan, R.D., Cobb, S., French, J.R.P., Van Harrison, R. and Pinneau, S.R. (1975) *Job Demands and Worker Health*. HEW publication (NIOSH).

Caplan, R.D., Cobb, S., French, J.R.P., Jr, Van Harrison R.V. and Pinneau, S.R. (1980) *Job Demands and Worker Health: Main Effects and Occupational Differences*. Ann Arbor, Michigan: Survey Research Centre, Institute of Social Research, University of Michigan.

Castles, S. (1985) *Immigrant Workers and Class Structure in Western Europe*. London: Oxford University Press.

Catalano, R., Rook, K. and Dooley, D. (1986) 'Labour markets and help-seeking: a test of the employment security hypothesis', *Journal of Health and Social Behaviour*, 27: 227–87.

Cecchini, P. (1988) *1992: The Benefits of a Single Market*. Aldershot: Wildwood House.

Chadwick, M. (1983) 'The recession and industrial relations: a factory approach', *Employee Relations*, 5: 5–12.

Chant, S.N. (1932) 'Measuring factors that make a job interesting', *Personnel Journal*, 11: 1–4.

Chinoy, E. (1955) *Automobile Workers and the American Dream*. Boston: Beacon.

Clegg, C. and Wall, T. (1981) 'Note on some new scales for measuring aspects of psychological well-being', *Journal of Occupational Psychology*, 54: 221–5.

Clegg, H. (1976) *Trade Unionism under Collective Bargaining*. Oxford: Blackwell.

Clock, T., Rosenkopf, L., Sambado, T. and Thompson, D. (1986) 'Layoffs at the Apple Computer Company: a case study'. Unpublished manuscript. Department of Industrial Engineering and Engineering Management, Stanford University.

Coch, L. and French, J.R.P. (1948) 'Overcoming resistance to change', *Human Relations*, 1: 512–32.

Cohen, F. and Lazarus, R.S. (1979) 'Coping with the stresses of illness', in G.C. Stone, F. Cohen and N.E. Adler (eds), *Health Psychology: A Handbook*. San Francisco: Jossey-Bass.

Cook, J. and Wall, T.D. (1980) 'New work attitude measures of trust, organizational commitment and personal need non-fulfilment', *Journal of Occupational Psychology*, 53: 39–52.

Cooper, W.H. and Richardson, A.J. (1986) 'Unfair comparisons', *Journal of Applied Psychology*, 71: 179–84.

Cornfield, D.B. (1983) 'Changes of layoff in a corporation: a case study', *Administrative Science Quarterly*, 28: 503–20.

Cottrell, L. (1942) 'The adjustment of the individual to his age and sex roles', *American Sociological Review*, 7: 617–20.

Cressey, P., Eldridge, J. and MacInnes, J. (1985) *Just Managing: Authority and*

Democracy in Industry. Milton Keynes: Open University Press.

Cyert, R.M. and March, J.G. (1963) *A Behavioral Theory of the Firm.* Englewood Cliffs, NJ: Prentice-Hall.

Daniel, W. and Millward, N. (1983) *Workplace Industrial Relations in Britain.* London: Heinemann.

D'Aunno, T. and Sutton, R.I. (1988) 'Organizational structure and job characteristics under financial threat: a partial test of the threat-rigidity thesis'. Paper presented at the Annual Meeting of the Academy of Management. Anaheim, Cal.

Davy, J.A., Kinicki, A.J., Scheck, C.L. and Sutton, C.L. (1988) 'Developing and testing an expanded model of survivor responses to layoffs: a longitudinal field study'. Paper presented at the Annual Meeting of the Academy of Management. Anaheim, Cal.

DeFrank, R.S. and Ivancevich, J.M. (1986) 'Job loss effects: an individual level model and review', *Journal of Vocational Behaviour*, 28: 1–20.

Depolo, M. and Sarchielli, G. (1985) 'Job insecurity, psychological well-being and social representation: a case of cost sharing'. Paper presented at the West European Conference on the Psychology of Work and Organization. Aachen, Federal Republic of Germany.

Dijkhuizen, N. van (1980) 'From stressor to strains: research into their interrelationship'. Dissertatie. Lisse: Swets en Zeitlinger.

Doeringer, P.B. and Piore, M.J. (1971) *Internal Labor Markets and Manpower Analysis.* Lexington, Mass.: D.C. Heath.

Donaldson, L. and Lynn, R. (1976) 'The conflict resolution process: the two factor theory and an industrial case', *Personnel Review*, 5: 21–8.

Donovan, V. (1968) Royal Commission on Trade Unions and Employers' Associations Report. Cmnd 3623. London: HMSO.

Dooley, D. and Catalano, R. (1980) 'Economic change as a behavioral disorder', *Psychological Bulletin*, 87: 450–68.

Dooley, D., Catalano, R. and Brownwell, A. (1986) 'The relation of economic conditions, social support and life events to depression', *Journal of Community Psychology*, 14: 103–19.

Downey, H.K., Hellriegel, D. and Slocum, J.W. (1975) 'Environmental uncertainty: the construct and its application', *Administrative Science Quarterly*, 20: 613–29.

Edmonds, J. (1984) 'Decline of the big battalions', *Personnel Management*, March: 18–21.

Edwards, C. and Heery, E. (1989) 'Recession in the public sector: industrial relations in Freightliner 1981–5', *British Journal of Industrial Relations*, 27: 57–71.

Edwards, P. (1985a) 'Managing labour relations through the recession', *Employee Relations*, 72: 3–7.

Edwards, P. (1985b) 'Managing the recession: the plant and the company', *Employee Relations*, 73: 4–8.

Edwards, P. (1985c) 'The myth of the macho manager', *Personnel Management*, 17: 32–5.

Eiser, J.R. (1987) *Social Psychology: Attitudes, Cognition and Social Behaviour.* Cambridge: Cambridge University Press.

Epstein, S. and Roupenian, A. (1970) 'Heart rate and skin conductance during experimentally induced anxiety: the effect of uncertainty about receiving a noxious stimulus', *Journal of Personality and Social Psychology*, 16: 20–8.

Erikson, K.T. (1976) *Everything in its Path: Destruction of Community in the Buffalo Creek Flood.* New York: Simon & Schuster.

Etzioni, A. (1961) *A Comparative Analysis of Complex Organizations.* New York: The Free Press.

Fagin, L. and Little, M. (1984) *The Forsaken Families.* Harmondsworth: Penguin.

Feather, N.T. (1982) *Expectations and Actions: Expectancy-Value Models in Psychology.* Hillsdale, NJ: Lawrence Erlbaum.

Feldstein, M. (1976) 'Temporary layoffs and the theory of unemployment', *Journal of Political Economy,* 84: 937–57.

Fink, L., Beak, J. and Taddeo, T. (1971) 'Organizational crisis and change', *Journal of Applied Behavioral Science,* 7: 15–37.

Ford, J.D. (1980) 'The administrative component in growing and declining organizations: a longitudinal analysis', *Academy of Management Journal,* 23: 615–30.

Fox, A. (1974) *Beyond Contract: Work, Power and Trust Relations.* London: Faber & Faber.

Fox, F.V. and Staw, B.M. (1979) 'The trapped administrators: effects of job insecurity and policy resistance upon commitment to a course of action', *Administrative Science Quarterly,* 24: 449–71.

Frankenhauser, M. (1981) 'Coping with stress at work', *International Journal of Health Services,* 11: 491–510.

Freeman, J. and Hannan, M.T. (1975) 'Growth and decline processes in organizations', *American Sociological Review,* 40: 215–28.

French, J.R.P., Jr, Caplan, R.D. and Van Harrison, R. (1982) *The Mechanisms of Job Stress and Strain.* New York: Wiley.

Fried, M. (1979) 'Role adaptation and the appraisal of work-related stress', in L.A. Ferman and J.P. Gordus (eds), *Mental Health and the Economy.* Kalamazoo, Mich.: W.E. Upjohn Institute for Employment Research.

Friedlander, F. and Walton, E. (1964) 'Positive and negative motivators toward work', *Administrative Science Quarterly,* 9: 194–207.

Fry, F.L. (1973) 'More on the causes of quits in manufacturing', *Monthly Labor Review,* June: 48–9.

Fryer, D. (1986) 'Employment deprivation and personal agency during unemployment', *Social Behaviour,* 1: 3–23.

Fryer, D. and Payne, R. (1986) 'Being unemployed: a review of the literature of the psychological experience of unemployment', in C.L. Cooper and I. Robertson (eds), *Review of Industrial and Organizational Psychology.* Chichester: Wiley.

Gannon, M.J., Foreman, C. and Pugh, K. (1973) 'The influence of a reduction in force on the attitudes of engineers', *Academy of Management Journal,* 16: 330–4.

Geddes, M. (1988) 'Social audits and social accounting in the UK: a review', *Regional Studies,* 22: 60–5.

Gennard, J. (1985) 'What's new in industrial relations?', *Personnel Management,* 17: 19–21, 37.

Gow, J.S., Clark, A.W. and Dossett, C.S. (1974) 'A path analysis of variables influencing labor turnover', *Human Relations,* 27: 703–19.

Graen, G.B. (1966) 'Motivator and hygiene dimensions for research and development engineers', *Journal of Applied Psychology,* 50: 563–6.

Greenhalgh, L. (1979) 'Job security and the disinvolvement syndrome: an exploration of patterns of worker behavior under conditions of anticipatory grieving over job loss'. PhD dissertation, Cornell University.

Greenhalgh, L. (1982) 'Maintaining organizational effectiveness during organizational retrenchment', *Journal of Applied Behavioral Science,* 18: 155–70.

Greenhalgh, L. (1983) 'Organizational decline', in S.B. Bacharach (ed.), *Research in the Sociology of Organizations,* vol. 2. Greenwich, Conn.: JAI Press.

Greenhalgh, L. (1985) 'Job insecurity and disinvolvement: field research on the survivors of layoffs'. Paper presented at the Annual Meeting of the Academy of Management, San Diego.

Greenhalgh, L. and Jick, T.D. (1979) 'The relationship between job insecurity and turnover, and its differential effects on employee quality level'. Paper presented at the Annual Meeting of the Academy of Management, Atlanta.

Greenhalgh, L. and Jick, T.D. (1983) 'The phenomenology of sense-making in a declining organization: effects of individual differences'. Paper presented at the Annual Meeting of the Academy of Management, Dallas.

Greenhalgh, L. and Kaestle, J.M. (1981) 'Severance pay in the public sector'. Paper presented at the Annual Meeting of the Academy of Management, San Diego.

Greenhalgh, L., Lawrence, A.T. and Sutton, R.I. (1988) 'The determinants of workforce reduction strategies in declining organizations', *Academy of Management Review*, 13: 241–54.

Greenhalgh, L. and McKersie, R.B. (1980) 'Cost effectiveness of alternative strategies for cutback management', *Public Administration Review*, 6: 575–84.

Greenhalgh, L., McKersie, R.B. and Gilkey, R.W. (1986) 'Rebalancing the workforce at IBM: a case study of redeployment and revitalization', *Organizational Dynamics*, 14: 30–47.

Greenhalgh, L. and Rosenblatt, Z. (1984) 'Job insecurity: towards conceptual clarity', *Academy of Management Review*, 9: 438–48.

Guest, D. (1982) 'Has the recession really hit personnel management?', *Personnel Management*, 14: 36–9.

Guest, D. and Fatchett, D. (1974) *Worker Participation: Individual Control and Performance*. London: Institute of Personnel Management.

Gutchess, J.F. (1985) *Employment Security in Action*. New York: Pergamon Press.

Hall, D.T. and Mansfield, R. (1971) 'Organizational and individual response to external stress', *Administrative Science Quarterly*, 16: 533–47.

Hanlon, M.D. (1979) 'Primary groups and unemployment'. PhD dissertation, Columbia University.

Hardy, C. (1985) *Managing Organizational Closure*. Aldershot: Gower.

Harris, S.E. and Sutton, R.I. (1986) 'Functions of parting ceremonies in dying organizations', *Academy of Management Journal*, 29: 5–30.

Hartley, J.F. (1980) 'The impact of employment upon the self-esteem of managers', *Journal of Occupational Psychology*, 53: 147–55.

Hartley, J.F. (1985) 'Job insecurity and union–management relations'. Paper presented at the West European Conference on the Psychology of Work and Organization. Aachen, Federal Republic of Germany.

Hartley, J.F. (1987) 'Unemployment: the wife's perspective and role', in S. Fineman (ed.), *Unemployment: Personal and Social Consequences*. London: Tavistock.

Hartley, J.F. (1988) 'Psychology and industrial relations: social processes in organizations', *International Journal of Comparative Labour Law and Industrial Relations*, 4: 53–60.

Hartley, J.F. and Cooper, C.L. (1976) 'Redundancy: a psychological problem?', *Personnel Review*, 5: 44–8.

Hartley, J.F. and Fryer, D. (1984) 'The psychology of unemployment: a critical appraisal', in G.M. Stephenson and J.H. Davis (eds), *Progress in Applied Social Research*, vol. 2. Chichester: Wiley.

Hartley, J.F. and Kelly, J. (1986) 'Psychology and industrial relations: from conflict to co-operation?', *Journal of Occupational Psychology*, 59: 161–76.

Hartley, J.F. and Klandermans, P.G. (1986) 'Individual and collective responses to job insecurity', in G. Debus and H.W. Schroif (eds), *The Psychology of Work and Organization*. Amsterdam: Elsevier Science Publishers.

Hartley, J. and Mohr, G. (1989) 'Arbeitsplatzverlust und Erwerblosigkeit', in S. Greif, H. Holling and N. Nicholson (eds), *Arbeits-und Organizationspsychologie*. Munich: Verlags Union.

Harvey, J.H. and Weary, G. (1985) *Attribution: Basic Issues and Application*. Orlando, Fla: Academic Press.

Hawkins, K. (1985) 'The "new realism" in British industrial relations?', *Employee Relations*, 7: 2–7.

Hayes, J. and Nutman, P. (1981) *Understanding the Unemployed*. London: Tavistock.

Heider, F. (1958) *The Psychology of Interpersonal Relations*. New York: Wiley.

Hersey, R.B. (1936) 'Psychology of workers', *Personnel Journal*, 8: 291–6.

Hershey, R. (1972) 'Effects of anticipated job loss on employee behaviour', *Journal of Applied Psychology*, 56: 273–5.

Herzberg, F. (1966) *Work and the Nature of Man*. Cleveland: World.

Herzberg, F. (1968) 'One more time: how do you motivate employees?', *Harvard Business Review*, 46 (Jan.–Feb.): 53–62.

Hewstone, M. (1988) 'Causal attribution: from cognitive processes to collective beliefs', *The Psychologist: Bulletin of the British Psychological Society*, 8: 323–7.

Hewstone, M. (1989) *Causal Attribution: From Cognitive Processes to Collective Beliefs*. Oxford: Blackwell.

Hewstone, M. and Jaspars, J.M.F. (1984) 'Social dimensions of attribution', in H. Tajfel (ed.), *The Social Dimension*. Cambridge: Cambridge University Press.

Hirschman, A.O. (1970) *Exit, Voice and Loyalty: Responses to Decline in Firms, Organizations and States*. Cambridge, Mass.: Harvard University Press.

Hirschorn, L. and associates (1983) *Cutting Back: Retrenchment and Redevelopment in Human and Community Services*. San Francisco: Jossey-Bass.

Hodson, R. and Kaufman, R. (1982) 'Economic dualism: a critical review', *American Sociological Review*, 47: 727–39.

Hoekstra, H.A. (1986) 'Cognition and affect in the appraisal of events'. Doctoral thesis. Groningen: Rijksuniversiteit Groningen.

Hovland, C. and Janis, J. (1959) *Personality and Persuasibility*. New Haven, Conn.: Yale University Press.

Hunter, L.C. (1980) 'The end of full employment?', *British Journal of Industrial Relations*, 18: 44–56.

Hurley, R.F., Grover, S., Read, T.F. and Brockner, J. (1988) 'Layoffs and survivors' reactions: a social information processing perspective'. Paper presented at the 1988 Academy of Management Meetings, Anaheim, Cal.

Hyman, R. (1975) *Industrial Relations: A Marxist Introduction*. London: Macmillan.

Hyman, R. (1987) 'Strategy or structure? Capital, labour and control', *Work, Employment and Society*, 1: 25–55.

Jackson, P.R. (1986) 'Toward a social psychology of unemployment: a commentary on Fryer, Jahoda, and Kelvin and Jarrett', *Social Behaviour*, 1: 33–9.

Jacobson, D. (1972) 'The influence of fatigue producing factors in industrial work on pre-retirement attitudes', *Occupational Psychology*, 46: 193–200.

Jacobson, D. (1985) 'Determinants of "job at risk" (JAR) behavior'. Paper presented at the West European Conference on the Psychology of Work and Organization. Aachen, Federal Republic of Germany.

Jacobson, D. (1987) 'A personological study of the job insecurity experience', *Social Behaviour*, 2: 143–55.

Jahoda, M. (1982) *Employment and Unemployment: A Social Psychological Analysis*. Cambridge: Cambridge University Press.

Janis, I.L. and Mann, L. (1977) *Decision Making*. New York: The Free Press.

Janney, J.G., Masuda, M. and Holmes, T.H. (1977) 'Impact of natural disasters on life events', *Journal of Human Stress*, 3: 22–3, 24–5.

Jick, T.D. (1979) 'Process and impacts of a merger: individual and organizational perspectives'. PhD dissertation, Cornell University.

Jick, T.D. and Greenhalgh, L. (1980) 'Realistic job previews: a reconceptualization'. Paper presented at the Annual Meeting of the Academy of Management, Detroit.

Jick, T.D. and Murray, V. (1982) 'The management of hard times: budget cutbacks in public sector organizations', *Organization Studies*, 3: 141–69.

Joelson, M. and Wahlquist, L. (1987) 'The psychological meaning of job insecurity and job loss: the results of a longitudinal study', *Social Science and Medicine*, 25: 179–82.

Johnson, C.D., Messe, L.A. and Crano, W.D. (1984) 'Predicting job performance of low income workers: the Work Opinion Questionnaire', *Personnel Psychology*, 37: 291–9.

Jurgensen, C.E. (1978) 'Job preferences (what makes a job good or bad)', *Journal of Applied Psychology*, 63: 267–76.

Kahn, R.L., Wolfe, D.M., Quinn, R.P., Snoek, J.D. and Rosenthal, R.A. (1964) *Organizational Stress*. New York: Wiley.

Kanabayashi, M. (1988) 'In Japan, employees are switching firms for better work, pay', *Wall Street Journal*, 11 Oct., A1, A19.

Kanter, R.M. (1985) *The Change Masters*. London: Unwin.

Karasek, R, (1979) 'Job demands, job decision latitude and mental strain: implications for job redesign', *Administrative Science Quarterly*, 24: 285–308.

Katz, D. and Kahn, R.L. (1978) *The Social Psychology of Organizations*. New York: Wiley.

Kaufman, H.G. (1982) *Professionals in Search of Work: Coping with the Stress of Job Loss and Unemployment*. New York: Wiley.

Kelley, H.H. and Michela, J.L. (1980) 'Attribution theory and research', *Annual Review of Psychology*, 31: 457–503.

Kelly, J. (1984) 'Management strategy and the reform of collective bargaining: cases from the British Steel Corporation', *British Journal of Industrial Relations*, 22: 135–53.

Kelly, J. (1987) 'Trade unions through the recession 1980–84', *British Journal of Industrial Relations*, 25: 275–82.

Kelly, J. and Bailey, R. (1989) 'British trade union membership, density, and decline in the 1980s', *Industrial Relations Journal*, 54–61.

Kelly, J. and Heery, E. (1989) 'Full-time officers and trade union recruitment', *British Journal of Industrial Relations*, 27: 196–213.

Kelly, J. and Kelly, C. (in press) '"Them and us": a social psychological analysis of the "new industrial relations"', *British Journal of Industrial Relations*.

Kelly, J. and Nicholson, N. (1980) 'Strikes and other forms of industrial action', *Industrial Relations Journal*, 11: 20–31.

Kelvin, P. and Jarrett, J. (1985) *Unemployment: Its Social Psychological Effects*. Cambridge: Cambridge University Press.

Kiechel, W. (1987) 'Your new employment contract', *Fortune*, 6 July, 81–2.

Kinicki, A.J. (1985) 'Personal consequences of plant closings: a model and preliminary test', *Human Relations*, 38: 197–212.

Kirkbride, P. (1985) 'Power in industrial relations research: a review of some recent work', *Industrial Relations Journal*, 16: 44–56.

Kissler, G.D. (1987) 'Smokestack transitions: managing phoenix of its dead ash'. Paper presented at the 1987 National Academy of Management Meetings, New Orleans, La.

Klandermans, P.G. (1983) *Participatie in een Sociale Beweging: Een Mobilizatiecampagne Onderzocht*. Amsterdam: VU-uitgeverij.

Klandermans, P.G. (1984a) 'Mobilization and participation in trade union action: a value expectancy approach', *Journal of Occupational Psychology*, 57: 107–20.

Klandermans, P.G. (1984b) 'Mobilization and participation: social psychological expansions of resource mobilization theory', *American Sociological Review*, 49: 583–600.

Klandermans, P.G. (1986) 'Psychology and trade union participation: joining, acting, quitting', *Journal of Occupational Psychology*, 59: 189–204.

Kleber, R.J. (1982) *Stressbenaderingen in de Psychologie*. Deventer: Van Lochum Slaterus.

Klein, H. (1974) 'Delayed effects and after-effects of severe traumatization', *Israel Annals of Psychiatry and Related Disciplines*, 12: 293–303.

Kobasa, S.C., Maddi, S.R. and Kahn, S. (1982) 'Hardiness and health: a prospective study', *Journal of Personality and Social Psychology*, 42: 168–77.

Kochan, T.A., Katz, H.C. and McKersie, R.B. (1986) *A Review of the Transformation of American Industrial Relations*. New York: Basic Books.

Kornhauser, A. (1965) *Mental Health of the Industrial Worker*. New York: Wiley.

Krantz, J. (1985) 'Group processes under conditions of organizational decline', *Journal of Applied Behavioral Science*, 21: 1–17.

Kressel, K. (1986) 'Patterns of coping in divorce', in R.H. Moos (ed.), *Coping with Life Crises*. New York: Plenum Press.

Kubler-Ross, E. (1969) *On Death and Dying*. New York: Macmillan.

Lahey, M.A. (1984) 'Job insecurity: its meaning and measure'. Unpublished doctoral dissertation, Kansas State University.

Lane, T. (1982) 'The unions: caught on the ebb tide', *Marxism Today*, September: 6–13.

Langer, E.J. and Rodin, J. (1976) 'The effects of choice and enhanced personal responsibility for the aged: a field experiment in an institutional setting', *Journal of Personality and Social Psychology*, 34: 191–8.

Lawler, E.E., 3rd and Hall, D.T. (1970) 'Relationship of job characteristics to job involvement, satisfaction, and intrinsic motivation', *Journal of Applied Psychology*, 54: 305–12.

Lawler, J. (1986) 'Union growth and decline: the impact of employer and union tactics', *Journal of Occupational Psychology*, 59: 217–30.

Lazarus, R.S. (1966) *Psychological Stress and the Coping Process*. New York: McGraw-Hill.

Lazarus, R.S. and Folkman, S. (1984) *Stress Appraisal and Coping*. New York: Springer.

Leadbetter, C. and Lloyd, J. (1987) *In Search of Work*. Harmondsworth: Penguin.

Leana, C.R. and Ivancevich, J.M. (1987) 'Involuntary job loss: institutional interventions and a research agenda', *Academy of Management Review*, 12: 301–12.

Lefcourt, H.M. (1976) *Locus of Control: Current Trends in Theory and Research*. Hillsdale, NJ: Lawrence Erlbaum.

Legge, K. (1988) 'Personnel management in recession and recovery: a comparative analysis of what the surveys say', *Personnel Review*, 17: 1–72.

Leighton, P. (1986) 'Marginal workers', in R. Lewis (ed.), *Labour Law in Britain*. Oxford: Blackwell.

Levine, C.H. (1979) 'More on cutback management: hard questions for hard times', *Public Administration Review*, 39: 179–83.

Levinson, H., Price, C.R., Munden, K.J., Mandl, H.J. and Solley, C.M. (1962) *Men, Management and Mental Health*. Cambridge, Mass.: Harvard University Press.

Lewin, K. (1951) *Field Theory in Social Science*. New York: Harper & Row.

Liberman, M.A. and Tobin, S.S. (1983) *The Experience of Old Age: Stress, Coping and Survival*. New York: Basic Books.

Lloyd, J. (1987) 'Can the unions survive?', *Personnel Management*, 19: 38–41.

Lodahl, T.M. and Kejner, M. (1965) 'The definition of measurement of job involvement', *Journal of Applied Psychology*, 49: 24–33.

Lomranz, J., Lubin, B., Eyal, N. and Medini, G. (1981) 'Hebrew version of the Depression Adjective Check List', *Journal of Personality Assessment*, 45: 380–4.

Louis, M.R. (1980a) 'Career transitions: varieties and commonalities', *Academy of Management Review*, 5: 329–40.

Louis, M.R. (1980b) 'Surprise and sense making: what newcomers experience in entering unfamiliar organizational settings', *Administrative Science Quarterly*, 25: 226–51.

Lubin, B. (1967) *Depression Adjective Check List: Manual*. San Diego: Educational and Industrial Service.

MacCrimmon, K.R. (1966) 'Descriptive and normative implications of the decision theory postulates', in K. Borch and J. Mossin (eds), *Risk and Uncertainty: Proceedings of a Conference Held by the International Economic Association*. New York: St Martins Press.

MacInnes, J. (1987) *Thatcherism at Work*. Milton Keynes: Open University Press.

March, J.G. and Simon, H.G. (1958) *Organizations*. New York: Wiley.

Marginson, P., Edwards, P., Martin, R., Purcell, J. and Sisson, K. (1988) *Beyond the Workplace: Managing Industrial Relations in Multi-plant Enterprises*. Oxford: Blackwell.

Marsh, C., Fraser, C. and Jobling, R. (1984) 'Political responses to unemployment'. Unpublished paper, presented to the ESRC workshop on employment and unemployment. London, April.

Martin, J. and Roberts, C. (1984) *Women and Employment: A Lifetime Perspective*. London: HMSO.

Martin, R. (1987) 'A new realism in industrial relations?', in S. Fineman (ed.), *Unemployment: Personal and Social Consequences*. London: Tavistock.

Maslow, A.H. (1943) 'A theory of human motivation', *Psychological Review*, 50: 370–96.

Maslow, A.H. (1954) *Motivation and Personality*. New York: Harper.

Massey, D. and Meegan, R. (1982) *The Anatomy of Job Loss*. London: Methuen.

McAdam, D., McCarthy, J.D. and Zald, M.N. (1988) 'Social Movements', in N.J. Smelser (ed.), *Handbook of Sociology*. Newbury Park: Sage, pp. 695–739.

McCaskey, M.B. (1976) 'Tolerance for ambiguity and the perception of environmental uncertainty in organizational design', in R. Kilmann, L. Pondy and D. Slevin (eds), *The Management of Organizational Design*. vol. 2. New York: Elsevier-North Holland.

McGregor, D. (1960) *The Human Side of Enterprise*. New York: McGraw-Hill.

McKelvey, B. (1982) *Organizational Systematics: Taxonomy, Evolution, Classification*. Berkeley: University of California Press.

McKersie, R.B., Greenhalgh, L. and Jick, T.D. (1981) 'The CEC: labor–management cooperation in New York', *Industrial Relations*, 20: 212–20.

McShane, S. (1986) 'The multidimensionality of union participation', *Journal of Occupational Psychology*, 59: 177–88.

Mechanic, D. (1978) *Medical Sociology*. New York: The Free Press.

Merton, R.K. (1962) 'The machine, the worker, and the engineer', in S. Nosow and W.H. Form (eds), *Man, Work and Society*. New York: Basic Books.

Merton, R.K. (1968) *Social Theory and Social Structure*. New York: The Free Press.

Metcalf, D. (1989) 'Water notes dry up: the impact of the Donovan reform proposals and Thatcherism at work on labour productivity in British manufacturing industry', *British Journal of Industrial Relations*, 27: 1–32.

Miles, R.M. and Randolph, W.A. (1980) 'Influence of organizational learning styles on early development', in J.R. Kimberly and R.H. Miles (eds), *The Organizational Life Cycle*. San Francisco: Jossey-Bass.

Milgram, N. (1986) *Stress and Coping in Time of War*. New York: Brunner-Mazel.

Milkovich, G.T., Anderson, J.C. and Greenhalgh, L. (1976) 'Organizational careers: environmental, organizational and individual determinants', in L. Dyer, (ed.), *Careers in Organizations: Individual Planning and Organizational Development*. Ithaca: New York State School of Industrial and Labor Relations, Cornell University.

Miller, S.M. (1981) 'Predictability and human stress: toward a clarification of evidence and theory', in L. Berkowitz (ed.), *Advances in Experimental Social Psychology*, 14. New York: Academic Press.

Milliken, F.J. (1987) 'Three types of perceived uncertainty about the environment: stage, effect, and response uncertainty', *Academy of Management Review*, 12: 133–43.

Millward, N. and Stevens, M. (1986) *British Workplace Industrial Relations 1980–84*. Aldershot, Gower.

Mitchell, T.R. (1982) 'Expectancy-value models in organizational psychology', in N.T. Feather (ed.), *Expectancy-Value Models in Psychology*. Hillsdale, NJ: Lawrence Erlbaum.

Moos, R.H. and Tsu, V.D. (1977) 'The crisis of physical illness: an overview', in R.H. Moos (ed.), *Coping with Physical Illness*. New York: Plenum.

Moscovici, S. (1981) 'On social representation', in J.P. Forgas (ed.), *Social Cognition: Perspectives on Everyday Understanding*. London: Academic Press.

Moscovici, S. (1984) 'The phenomenon of social representations', in R. Farr and S. Moscovici (eds), *Social Representations*. Cambridge: Cambridge University Press.

Moscovici, S. and Hewstone, M. (1983) 'Social representations and social explanations: from the "naive" to the "amateur" scientist', in M. Hewstone (ed.), *Attribution Theory: Social and Functional Extensions*. Oxford: Basil Blackwell.

Mowday, R.T., Porter, L.W. and Steers, R.M. (1982) *Employee–Organization Linkages: The Psychology of Commitment, Absenteeism, and Turnover*. New York: Academic Press.

Nash, M. (1983) *Managing Organizational Performance*. San Francisco: Jossey-Bass.

Nicholson, N. (1976) 'The role of the shop steward', *Industrial Relations Journal*, 7: 15–26.

Nicholson, N. (1984) 'A theory of work role transitions', *Administrative Science Quarterly*, 29: 172–91.

Nicholson, N., Ursell, G. and Blyton, P. (1981) *The Dynamics of White Collar Unionism*. London: Academic Press.

Nicholson, N. and West, M. (1988) *Managerial Job Change: Men and Women in Transition*. Cambridge: Cambridge University Press.

Nickson, A. and Gaffakin, F. (1984) *Jobs Crisis and the Multi-Nationals: Deindustrialization in the West Midlands*. Birmingham: Trade Union Resource Centre.

Northcott, J., Fogarty, M. and Trevor, M. (1985) *Chips and Jobs: Acceptance of New Technology at Work*. London: Policy Studies Institute.

O'Brien, G.E. (1986) *Psychology of Work and Unemployment*. Chichester: Wiley.

OECD (1985a) 'Employment in small and large firms: where have the jobs come from', *Employment Outlook*. Paris: OECD.

OECD (1985b) 'Employment growth, flexibility and job security: a challenge for all', *Employment Outlook*. Paris: OECD.

Ogden, S. (1981) 'The reform of collective bargaining: a managerial revolution?', *Industrial Relations Journal*, 12: 30–42.

Oliver, P.E. and Marwell, G. (1988) 'Mobilizing technologies and the micro-meso link'. Paper presented at the Workshop on Frontiers in Social Movement Theory, University of Michigan, Ann Arbor.

Parkes, C.M. (1972) *Bereavement*. London: Tavistock.

Parsons, D. (1972) 'Specific human capital: an application to quit rates and layoff rates', *Journal of Political Economy*, 80: 1120–43.

Paulay, J.D. (1986) 'Slow death: one survivor's experience', in R.H. Moos (ed.), *Coping with Life Crises*. New York: Plenum Press.

Pennings, J.M. (1975) 'The relevance of the structural-contingency model for organizational effectiveness', *Administrative Science Quarterly*, 20: 393–410.

Pettigrew, A. (1983) 'Patterns of managerial response as organizations move from rich environments to poor environments', *Educational Management and Administration*, 11: 104–14.

Pfeffer, J. (1981) *Power in Organizations*. Marshfield, Mass.: Pitman.

Pfeffer, J. (1983) 'Organizational demography', in L.L. Cummings and B.M. Staw (eds), *Research in Organizational Behaviour*, vol. 5. Greenwich, Conn.: JAI Press.

Pfeffer, J. and Baron, J.N. (1988) 'Taking the workers out: recent trends in the structuring of employment', in B.M. Staw and L.L. Cummings (eds), *Research in Organizational Behaviour*, vol. 10. Greenwich, Conn.: JAI Press.

Porter, L.W. and Lawler, E.E. (1968) *Managerial Attitudes and Performance*. Homewood, Ill.: Irwin-Dorsey.

Porter, L.W., Steers, R.M., Mowday, R.T. and Boulian, P.V. (1974) 'Organisational commitment, job satisfaction, and turnover among psychiatric technicians', *Journal of Applied Psychology*, 59: 603–9.

Price, R. and Bain, G. (1983) 'Union growth in Britain: retrospect and prospect', *British Journal of Industrial Relations*, 21: 46–68.

Price, J.L. (1977) *The Study of Turnover*. Armes: Iowa State University Press.

Purcell, J. (1981) *Good Industrial Relations*. London: Macmillan.

Purcell, J. and Sisson, K. (1983) 'Strategies and practice in the management of industrial relations', in G. Bain (ed.), *Industrial Relations in Britain*. Oxford: Blackwell.

Quinn, J.F. (1979) 'Wage determination and discrimination among older workers', *Journal of Gerontology*, 35: 728–35.

Quinn, R.P. (1973) *Locking-in as a Moderator for the Relationship Between Job Satisfaction and Mental Health*. Ann Arbor: Survey Research Centre, University of Michigan.

Rabinowitz, S. and Hall, D.T. (1977) 'Organizational research on job involvement', *Psychological Bulletin*, 84: 265–88.

Roethlisberger, F.G. and Dickson, W.J. (1946) *Management and the Worker*. Cambridge, Mass.: Harvard University Press.

Rogers, R. (1985) *Guests Come to Stay: The Effects of European Labor Migration on Sending and Receiving Countries*. New York: Westview.

Rose, M. and Jones, B. (1986) 'Redividing labour: factor politics and work reorganization in the current industrial transition', in K. Purcell, S. Wood, A. Watson and S. Allen (eds), *The Changing Experience of Employment: Restructuring and Recession*. Basingstoke: Macmillan.

Rosow, J.M. and Zager, R. (1984) *Employment Security in a Free Economy*. New York: Pergamon Press.

Rothman, R.A., Schwartzbaum, A.M. and McGrath, J.H., 3rd (1971) 'Physicians and a hospital merger: patterns of resistance to organizational change', *Journal of Health and Social Behavior*, 12: 46–55.

Rotter, J.B., Chance, J.E. and Phares, E.J. (1972) *Applications of a Social Learning Theory of Personality*. New York: Holt, Rinehart & Winston.

Schaufeli, W.B. (1988) 'Unemployment and psychological health: an investigation under Dutch professionals'. Doctoral thesis. Groningen: Rijksuniversiteit, Groningen.

Schein, E.H. (1968) 'Organizational Socialization', *Industrial Management Review*, 2: 37–45.

Schein, E.H. (1971) 'The individual, the organization, and the career: a conceptual scheme', *Journal of Applied Behavioral Science*, 7: 401–26.

Schein, E.H. (1980) *Organizational Psychology* (3rd edn). Englewood Cliffs, NJ: Prentice-Hall.

Schein, V. (1979) 'Examining an illusion: the role of deceptive behaviors in organizations', *Human Relations*, 32: 287–95.

Schlossberg, N. and Liebowitz, A. (1980) 'Organizational support systems as buffers to job loss', *Journal of Vocational Behavior*, 17: 401–26.

Schuler, R.S. (1975) 'Sex, organizational level and outcome importance: where the differences are', *Personnel Psychology*, 28: 365–75.

Schwartz, H.A. (1980) 'Job involvement, job enrichment and obsession–compulsion'. Paper presented at the annual meeting of the Eastern Academy of Management, Buffalo.

Seashore, S. and Yuchtman, E. (1967) 'A system resource approach to organizational effectiveness', *American Sociological Review*, 32: 891–903.

Seligman, M.E.P. (1975) *Helplessness*. San Francisco: W.H. Freeman.

Semyonov, M. and Levin-Epstein, N. (1987) *Hewers of Wood and Drawers of Water: Noncitizen Arabs in the Israeli Labor Market*. Ithaca, NY: ILR Press.

Sherif, M. (1966) *Group Conflict and Cooperation*. London: Routledge & Kegan Paul.

Sims, J.H. and Baumann, D.D. (1974) 'The tornado threat: coping styles in the North and South', in J.H. Sims and D.D. Baumann (eds), *Human Behavior and the Environment: Interactions Between Man and his Physical World*. Chicago: Maaroufa Press.

Slote, A. (1969) *Termination at Baker Plant*. Indianapolis: Bobbs-Merrill.

Smart, C. and Vertinsky, I. (1977) 'Designs for crisis decision units', *Administrative Science Quarterly*, 22: 640–57.

Smith, D.T. (1974) *Racial Disadvantage in Employment*. London: PEP.

Sokol, M. and Louis, M.R. (1984) 'Career transitions and life event adaptation: integrating alternative perspectives on role transitions', in V.L. Allen and E. Van de Vliert (eds), *Role Transitions*. New York: Plenum Press.

Spencer, B. (1985) 'Shop steward resistance in the recession', *Employee Relations*, 7: 22–8.

Staw, B.M., Sandelands, L.E. and Dutton, J.E. (1981) 'Threat-rigidity effects in organizational behavior: a multilevel analysis', *Administrative Science Quarterly*, 26: 501–24.

Stoikow, V. and Raimon, R.L. (1968) 'Determinants of differences in the quit rate among industries', *American Economic Review*, 58: 1283–98.

Strange, W. (1977) 'Job loss: a psychological study of worker reaction to a plant closing in a company town in Southern Appalacia'. Unpublished doctoral dissertation, Cornell University.

Strauss, G. (1984) 'Industrial relations: time of change', *Industrial Relations*, 23: 1–15.

Streeck, W. (1985) 'Industrial relations and industrial change in the motor industry: an international view'. Public lecture. University of Warwick: Industrial Relations Research Unit.

Sugarman, M. (1978) *Employer Dualism in Personnel Policies and Practices: Its Labor Turnover Implications*. Report prepared for Region IX, Employment and Training Administration, US Department of Labor.

Sutton, R.I. (1983) 'Managing organizational death', *Human Resource Management*, 22: 391–412.

Sutton, R.I. (1984) 'Organizational death'. PhD dissertation, University of Michigan.

Sutton, R.I. (1987) 'The process of organizational death: disbanding and reconnecting', *Administrative Science Quarterly*, 32: 542–69.

Sutton, R.I., Eisenhardt, K.M. and Jucker, J.V. (1986) 'Managing organizational decline: lessons from Atari', *Organizational Dynamics*, Spring: 17–29.

Sutton, R.I. and Kahn, R.L. (1987) 'Prediction, understanding, and control as antidotes to organizational stress', in J. Lorsch (ed.), *Handbook of Organizational Behaviour*. Englewood Cliffs, NJ: Prentice-Hall.

Sutton, R.I. and Schurman, S.J. (1985) 'On studying emotionally hot topics: lessons from an investigation of organizational death', in D.N. Berg and K.K. Smith (eds), *Exploring Clinical Methods for Social Research.* Beverly Hills: Sage.

Swinburne, P. (1981) 'The psychological impact of unemployment on managers and professional staff', *Journal of Occupational Psychology,* 54: 47–64.

Sykes, A.J.M. (1965) 'Myth and attitude change', *Human Relations,* 18: 323–37.

Tajfel, H. (1975) 'The exit of social mobility and the voice of social change: notes on the social psychology of intergroup relations', *Social Science Information,* 14: 101–18.

Taubman, P. and Wachter, M.L. (1986) 'Segmented labor markets', in O. Ashenfelter and R. Layard (eds), *Handbook of Labor Economics.* Amsterdam: Elsevier Science Publishers.

Taylor, S.E. and Brown, J.D. (1988) 'Illusion and well-being: a social psychological perspective on health', *Psychological Bulletin,* 103: 193–210.

Taylor, J.A. and Spence, K.W. (1952) 'The relationship of anxiety level to performance in serial learning', *Journal of Experimental Psychology,* 44: 61–4.

Terry, M. (1983) 'Shop steward development and managerial strategies', in G. Bain (ed.), *Industrial Relations in Britain.* Oxford: Blackwell.

Terry, M. (1986) 'How do we know if shop stewards are getting weaker?', *British Journal of Industrial Relations,* 24: 169–79.

Terry, M. and Edwards, P. (1988) *Shopfloor Politics and Job Controls.* Oxford: Blackwell.

Thayer, L. (1967) 'Communication and organization theory', in F.E.X. Dance (ed.), *Human Communication Theory.* New York: Holt, Rinehart & Winston.

Thompson, J.D. (1967) *Organizations in Action.* New York: McGraw-Hill.

Time Magazine (1987) 'Rebuilding to survive', 16 February: 36–40.

Time Magazine (1989) 'Where's the gung-ho?', 18 September: 26–8.

Towers, B. (1982) 'The economy, unemployment and industrial relations', *Industrial Relations Journal,* 13: 7–12.

Tsouderos, J.E. (1955) 'Organizational change in terms of a series of selected variables', *American Sociological Review,* 20: 206–10.

US Commission on Civil Rights (1977) *Last Hired, First Fired: Layoffs and Civil Rights.* Washington, DC: US Government Printing Office.

Van Maanen, J. (1976) 'Breaking in: socialization to work', in R. Dubin (ed.), *Handbook of Work, Organization and Society.* Chicago: Rand-McNally.

Van Maanen, J. and Schein, E.H. (1979) 'Toward a theory of organizational socialization', in B.M. Staw (ed.), *Research in Organizational Behavior,* vol. 1. Greenwich, Conn.: JAI Press.

Van Rooijen, L. (1984) 'Depressiegevoelens onder de Gewone Bevolking. Onderzoeksmemorandum RM-PS-84-18'. Amsterdam: Vakgroep Sociale Psychologie, Vrije Universiteit.

Veen, G., Van der, and Klandermans, P.G. (in press) 'Changes in the action strategies of trade unions'. Paper to the 4th West European Conference on the Psychology of Work and Organization. Cambridge: March.

Vroom, V.H. (1964) *Work and Motivation.* New York: Wiley.

Waddington D.P. (1986) 'The Ansell Brewery dispute: a social-cognitive approach to the study of strikes', *Journal of Occupational Psychology,* 59: 231–46.

Wallimann, I. (1984) 'The import of foreign workers in Switzerland: labor reproduction costs, ethnic antagonism, and the integration of foreign workers into Swiss society', *Research in Social Movements, Conflict and Change,* 7: 153–75.

Warr, P. (1978) 'A study of psychological well-being', *Journal of Occupational Psychology,* 69: 111–21.

Warr, P.B. (1984) 'Job loss, unemployment and well-being', in V.L. Allen and E. Van de Vliert (eds), *Role Transitions*, New York: Plenum Press.

Warr, P.B. (1987) *Work, Unemployment and Mental Health*. Oxford: Oxford University Press.

Warr, P.B., Cook, J. and Wall, T.D. (1979) 'Scales for the measurement of some work attitudes and aspects of psychological well-being', *Journal of Occupational Psychology*, 52: 129–48.

Weber, M. (1947) *The Theory of Social and Economic Organization* (trans. T. Parsons and A. Henderson). New York: Free Press.

Weick, K.E. (1979) *The Social Psychology of Organizing* (2nd edn). Reading, Mass.: Addison-Wesley.

Weiner, B. (1985) '"Spontaneous" causal thinking', *Psychological Bulletin*, 97: 74–84.

Whetten, D.A. (1979) 'Organizational responses to scarcity: difficult choices for difficult times'. Working paper, College of Commerce and Business Administration, University of Illinois at Urbana-Champaign.

Whetten, D.A. (1980a) 'Organizational decline: a neglected topic in organizational science', *Academy of Management Review*, 5: 577–88.

Whetten, D.A. (1980b) 'Organizational decline: sources, responses, and effects of', in J.R. Kimberly and R.H. Miles (eds), *The Organizational Life Cycle*. San Francisco: Jossey-Bass.

Winchester, D. (1981) 'Trade unions and the recession', *Marxism Today*, September: 20–5.

Winchester, D. (1983) 'Industrial relations research in Britain'. *British Journal of Industrial Relations*, 21: 100–14.

Winnubst, J.A.M. (1980) 'Stress in Organisaties: Naar een Nieuwe Benadering van Werk en Gezondheid', in P.J.D. Drenth, H. Thierry, P.J. Willems and C.J. de Wolff (eds), *Handboek Arbeids- en Organisatiepsychologie*. Deventer: Van Loghum Slaterus.

Wong, P.T.P. and Weiner, B. (1981) 'When people ask "why" questions, and the heuristics of attributional search', *Journal of Personality and Social Psychology*, 40: 650–63.

Zammuto, R.F. and Cameron, K. (1985) 'Environmental decline and organizational response', in L.L. Cummings and B.M. Staw (eds), *Research in Organizational Behavior*. Greenwich, Conn.: JAI Press.

Index

Note: Page references in italics indicate tables and figures.

Index compiled by Meg Davies (Society of Indexers)